T0181310

Well-Being Factors for Different Industries

Werner Seiferlein

Well-Being Factors for Different Industries

Tools for Determination

Mirjam Becker, Editor for the Author

Werner Seiferlein
TIM Technologie-Innovation-Management
Frankfurt/Main, Hessen
Germany

ISBN 978-3-658-34999-8 ISBN 978-3-658-34997-4 (eBook)
https://doi.org/10.1007/978-3-658-34997-4

The translation was done with the help of artificial intelligence (machine translation by the service DeepL.com). A subsequent human revision was done primarily in terms of content.

This Springer imprint is published by the registered company Springer Fachmedien Wiesbaden GmbH, part of Springer Nature.
The registered company address is: Abraham-Lincoln-Str. 46, 65189 Wiesbaden, Germany

Foreword

"Health is something very personal"—Only those who feel well are healthy, and those who are healthy feel well. This contrasts with the frequently expressed opinion: "You can feel good at home and on holiday. In production, what counts is effectiveness and efficiency. People are taken into account by considering ergonomics and, if necessary, in adapted production processes."

In times of Industry 4.0, humans—as it is felt—take a back seat to the "all-capable" software/machine. In this situation, is it worth taking a deeper look at "well-being factors for different industries in production"? The answer must clearly be: "Yes."

The human being is and remains—especially in times of Industry 4.0—the most important "production medium." Maintaining and expanding their necessary competence and experience as well as their flexibility, activity, and insensitivity to stress require active support from their environment. Industrie 4.0 provides suitable tools and working environments for this purpose. In the best case, this active support takes place over the entire life cycle of a "production" and therefore sensibly begins with the "design of the production" (building and production plant). As a result, the human factor "well-being" must be taken into account in addition to the usual production planning factors.

If "well-being/health" is something very personal, how can one then find and define "well-being factors" in planning that represent this "personal" feeling and make it manageable? Are there well-being factors that exist across sectors and, if so, which ones?

Our own experience shows that well-being factors, their effects, costs, and implementation in real plants of the process industry are often thought about in the planning/design phase, but there is still a lack of systematic theoretical elaboration and implementation.

In the planning of office areas, for example, well-being factors have long been systematically developed and taken into account. It has been shown that users must be closely involved in the planning process. The design of workspaces, the materials used, physical influences (noise, air, light), and psychological influences (colours, contact possibilities, privacy, furnishing of the workplace) but also organizational measures and identification with the corporate philosophy (and its implementation) have an influence on the well-being of an employee. Under the premise that the life cycle of such a facility is several decades, it makes sense to regularly review and adjust the well-being requirements to safeguard the investment, because these can change with the generations.

"Feel-good"elements generate costs in implementation (at first), then benefits in application, and when applied appropriately, are profitable for both the company and the employee.

Michael Müller
Head of Operations and Project Management
DI PA SE&C Engineering & Consulting
Munich, Germany

Niko Stantis
Head of Sales
DI PA SE&C Engineering & Consulting
Munich, Germany

Preface

Health is one of the megatrends of our time. The World Health Organization defines three areas as crucial for health: mental, social, and physical well-being (Fig. 1).

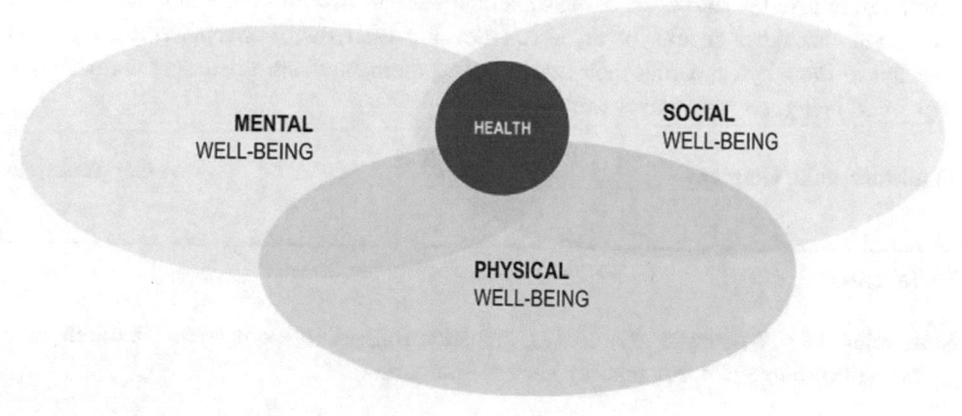

Fig. 1 The three factors influencing health

Also in relation to work and the working environment, all 3 levels play an essential role in human satisfaction.[1] When we think of the word "well-being," we all think of similar associations, some of which do not seem to translate to the world of work, such as rest and relaxation. In this book, well-being is seen as a heading to a puzzle picture, with each well-being factor researched or found empirically representing a piece of the puzzle. With the puzzle picture assembled, one is then able to create individual puzzle pictures for any industry.

The research on well-being factors has shown that many industries or fields are concerned with the factors that are very much influenced by human beings. From the

[1] Cf. Seiferlein, W. and Kohlert, C. (2018), pp. XXVf.

relevant literature, it can be seen that well-being indications have developed in different industries, which can act as a motivator to see if these observations correspond to the reality in other industries and fields. Indeed, in the present research, the application of well-being factors could be found in all of the selected industries and sectors. However, if there is no opportunity to think outside the box, the industries and sectors remain to themselves. With the active beginning of an exchange of information does it become clear that similar factors are known in them. Of course, from branch to branch there are additions of factors that are also applicable in other fields. Thus, it becomes meaningless from which branches and areas the found factors originate. The well-being factors examined in this thesis were determined with specialists in the course of interviews or discussions. The information is summarized in a kind of conclusion at the end of each chapter, and where necessary and possible, a recommendation for action is included.

The aim is to map the possible well-being factors and communication flows for the corresponding sectors or areas in planning and design terms. For this purpose, the factors found are transferred into a target matrix and applied to two areas as examples. It is then envisaged that other sectors or areas can develop their own concept with the factors relevant to them by mirroring their own user requirements from this target matrix, i.e. a structured feel-good matrix as it were.

Frankfurt/Main, Germany Werner Seiferlein

Reference

Seiferlein, W.; Woyczyk, R. (2017), Projekterfolg—Die vernetzten Faktoren von Investitionsprojekten, Fraunhofer Verlag, Stuttgart.

Contents

About the Author

Werner Seiferlein • Experienced as plant engineer in the field of drug production and process development (synthesis and biochemistry), pharmaceutical manufacturing (solids and liquids), as well as product development

- Carrying out site assessments (Moscow, Warsaw, and Kiev)
- Preparation of due diligences for the acquisition of pharmaceutical companies or the closing of alliances (e.g. planning of sterile production in the USA)
- Know-how in the field of FM (pharmaceutical energies, maintenance excellence, etc.)
- Know-how in real estate management (leasing, conversion, planning of new buildings, relocation of office buildings, etc.)
- Preparation and continuation of site master plans (Kansas City, Kawagoe, Suzano, and Frankfurt)
- International experience with projects for new buildings and conversions worldwide (Egypt, India, USA, Japan, etc.)
- Contacts with pharmaceutical companies via ISPE DACH (pharmaceutical association for engineers, committee member since 2005)
- Project management knowledge from numerous projects and preparation of a doctorate on this topic
- Honorary professorship at the Frankfurt University of Applied Sciences (2013)
- Management experience in the context of group and department head activities
- ISPE, active in the extended executive board and Fraunhofer steering committee member IAO
- DIN Standards Committee

Abbreviations

4 L	Light, air, noise, body
AI	Artificial intelligence
API	Active pharmaceutical ingredients
APV	Pharmacists Association
ART	Attention restoration theory
CAD	Computer-aided design
CEO	Chief executive officer
CIC	Clariant Innovation Center
CLIA	Cruise Lines International Association
DACH	Germany, Austria, Switzerland
DB	Deutsche Bundesbahn
EHS	Environment, health, and safety
EKZ	Shopping centre
ELA	Electro-acoustic system
EUniWell	European University for Well-Being
F&E	Research and development
ISPE	International Society for Pharmaceutical Engineering
KPI	Key performance indicator
NPV	Net present value
PAT	Process analytical technology
SB	Self-service

List of Figures

Introduction: Future 4.0

1

Preliminary Remarks

Globalization, consolidation, digitization: For many years, the vision of imminent change has been described, welcomed, lamented or feared. So far, the focus of the discussion has mostly been on the expected shifts in competition: relocation of production, outsourcing, productivity increases and cost reductions have been central terms within the debate. Now, however, the "new world" is reaching. Digitalization will make markets and companies more effective and efficient by radically change of the way we work and live. Thus, both individual workers and society as a whole are affected by these changes.[1]

The society benefits from this development towards the ever better or different. However, this development also has its dark sides. The development "towards the better"always presupposes a continuous change, which, however, often overwhelmed us.

Further Remarks: Future 4.0

Where in the past there was *one* solution for *one* application, today there are many more. For example, in the past there was a maximum of one ladies' and one men's bike, but today there are mountain bikes, trekking bikes, racing bikes, city bikes, e-bikes or pedelecs, children's or youth bikes, cross bikes, BMX bikes, folding or collapsible bikes. In order to make the right choice, methodical approaches must be taken and a plan drawn up. Something similar is happening in the automotive industry. Everyone talks about standards, which makes you think that diversity is being reduced to a few standardized car types. On the contrary, however, individualization is advancing with the system of "mass customization". Changes in this direction have the positive aspect that with increasing diversity, needs can be met more precisely, which is good for the customer.

[1] Cf. Seiferlein and Kohlert (2018).

W. Seiferlein, *Well-Being Factors for Different Industries*,
https://doi.org/10.1007/978-3-658-34997-4_1

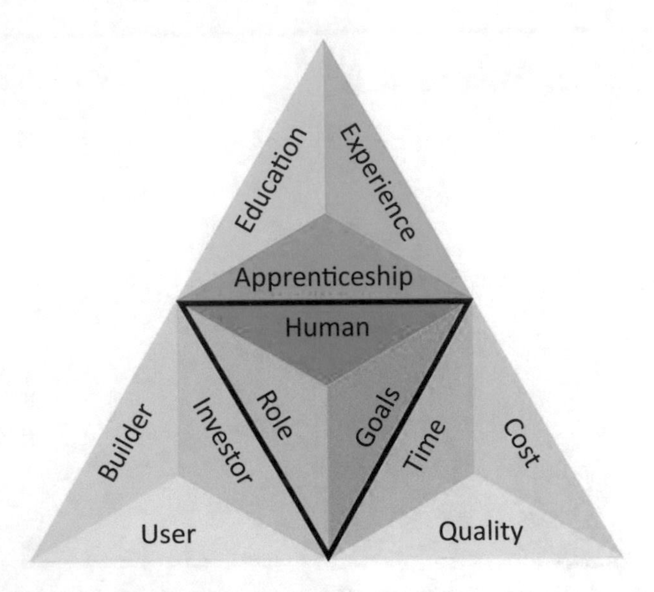

Fig. 1.1 The "networked triangle" and its factors

The conceptualisation of the office world has also changed dramatically over the last 10 to 15 years, ranging from the traditional cubicle office to the group, team or combination office, to the open space office, none territorial office and other combinations, e.g., combination office with shared desk.

The human factor influences the interaction between the actors in numerous areas, such as the processing of projects. This situation is illustrated in the **networked triangle**[2] and its factors or associated triangles.

The emotional triangle refers to what makes up a person in the working world: his training, his upbringing and his experience. The hierarchical triangle encompasses the role that everyone plays in a project, as they can be either investor, builder or operator of an object. The third is the magic triangle of success, long known from project management, which relates the factors of time, cost and quality (Fig. 1.1).

If a project is to lead to success and the satisfaction of all those involved, all factors of the networked triangle must be in balance. If, for example, a company realigns itself strategically, this has a direct impact on the employees and production and the measures must be supported equally by all those involved. Satisfaction with the working environment is significantly linked to the individual or collective perception and evaluation of a change process. Therefore, it is advantageous to integrate the user requirements of the employees into the office planning. This requires extensive information about the possibilities for participation in the process.

[2]Cf. Seiferlein and Woyczyk (2017), p. 16.

Change management supports the willingness of employees to change, so that they learn to understand that not only their company, but above all they themselves will undertake a transformation. People play the biggest role in this process and have a significant influence on the structure, work locations, focus of activities, types of work, demographics, etc. In this context, it is important that the orientation of the employees as well as that of the company is carried out in step. In this context, management bears a great responsibility, not only in deciding on essential points, but also with regard to its role as a role model. It is a no-go to ask employees to do something that management itself does not want to do—when the world turns, it turns for everyone. Therefore, as part of the change process, an Employees & Company Alignment[3] must be incorporated in advance and communicated accordingly. Another aspect concerns employee engagement, which increases in proportion to job satisfaction.[4]

> The production cycle of immaterial work will no longer be defined by the four walls of the factory. Thus, in terms of time and space, it will become increasingly difficult to distinguish between work and leisure. In a certain sense, life will be inseparable from work.[5]

It follows that well-being, as well as health and quality improvement, occupy a central position. In this context, it is important to define the goal and answer the question: "What is the benefit of introducing well-being factors?" Singular measures that offer disjointed outcomes depending on the request are seen as a patchwork and are usually poorly thought through. Furthermore, it also makes no sense to identify an object with all the different factors identified so that wellbeing factors are applied by hook or by crook. It seems important to raise awareness in order to obtain an overall picture of the object. In doing so, an iterative approach can ultimately be used to develop an overall picture, i.e. a master plan.

With the application of well-being factors, the performance of the person improves and an optimisation and improvement of the physical, mental and social competences sets in. This also reduces absenteeism from work and fluctuation. Strengthening well-being, creating added value, providing a defined working atmosphere and ensuring the maintenance of health are of paramount importance.

With such a measure, a company can increase its attractiveness, its entrepreneurial spirit and its spirit of research, because it positively stimulates the motivation of employees and creates an atmosphere that can withstand the "war for talents", thereby supporting innovations that ultimately increase production output.

If a project is to lead to success and the satisfaction of all those involved, all factors of the networked triangle must be in balance. If, for example, a company realigns itself strategically, this has a direct impact on the employees and production and the measures must be supported equally by all those involved. Satisfaction with the working

---

[3] See Sect. 5.3.2.

[4] Cf. Steelcase (2016), p. 7.

[5] Rumpfhuber (2013), p. 18.

environment is significantly linked to the individual or collective perception and evaluation of a change process. Therefore, it is advantageous to integrate the user requirements of the employees into the office planning. This requires extensive information about the possibilities for participation in the process.

Change management supports the willingness of employees to change, so that they learn to understand that not only their company, but above all they themselves will undergo a transformation. People play the biggest role in this process and have a significant influence on the structure, work locations, focus of activities, types of work, demographics, etc. In this context, it is important that the orientation of the employees as well as that of the company is carried out in step. In this context, management bears a great responsibility, not only in deciding on essential points, but also with regard to its role as a role model. It is a no-go to ask employees to do something that management itself does not want to do—when the world turns, it turns for everyone. Therefore, as part of the change process, an Employees & Company Alignment[6] must be incorporated in advance and communicated accordingly. Another aspect concerns employee engagement, which increases in proportion to job satisfaction.[7]

> The production cycle of immaterial work will no longer be defined by the four walls of the factory. Thus, in terms of time and space, it will become increasingly difficult to distinguish between work and leisure. In a certain sense, life will be inseparable from work.[8]

It follows that well-being, as well as health and quality improvement, occupy a central position. In this context, it is important to define the goal and answer the question: "What is the benefit of introducing well-being factors?" Singular measures that offer disjointed outcomes depending on the request are seen as a patchwork and are usually poorly thought through. Furthermore, it also makes no sense to identify an object with all the different factors identified so that wellbeing factors are applied by hook or by crook. It seems important to raise awareness in order to obtain an overall picture of the object. In doing so, an iterative approach can ultimately be used to develop an overall picture, i.e., a master plan.

With the application of well-being factors, the performance of the person improves and an optimisation and improvement of the physical, mental and social competences sets in. This also reduces absenteeism from work and fluctuation. Strengthening well-being, creating added value, providing a defined working atmosphere and ensuring the maintenance of health are of paramount importance.

With such a measure, a company can increase its attractiveness, its entrepreneurial spirit and its spirit of research, because it positively stimulates the motivation of employees and creates an atmosphere that can withstand the "war for talents", thereby supporting innovations that ultimately increase production output.

---

[6] See Sect. 5.3.2.

[7] Cf. Steelcase (2016), p. 7.

[8] Rumpfhuber (2013), p. 18.

1.1 Employees

In Germany, approximately 44.9 million people are employed (as of 2018/2019). In a simple classification of economic activities, only three areas can be distinguished:

- the first area covers agriculture, forestry and fisheries,
- the second sector is the manufacturing industry and
- the third area the service sector.

In 1970, 8.4% of all employed persons were still working in the first sector, in 1990 the share was 3.5% and in 2012 it was only 1.6%. The second sector has also become less important: in 1970, 46.5% of the workforce were employed in the manufacturing sector, but by 1990 this share had fallen steadily to 36.6%. In 2012, only every fourth person in employment was still working in the second sector (26%). In line with these developments, the importance of the third sector has steadily increased. The share of employees working in this sector increased from 45.1% in 1970 to 59.9% in 1990 to 73.7% in 2012. In the data, it should also be noted that the distribution across the different sectors cannot always be equated with the actual activity of an employee. For example, employees in the manufacturing sector do not necessarily work in actual production, but can also work in the commercial department of an industrial company.[9]

The age composition of employees has never been as heterogeneous as it is today, so that three generations usually work together in one team. In order to be able to serve the different needs of these groups, a general rethinking of work organisation is necessary. However, this also results in more insecure employment relationships and a splitting up of employees, e.g., due to outsourcing.[10] Another aggravating factor is the fact that the baby boomers will soon be retiring.[11] This raise concerns as to whether the number of retiring workers can be filled by potential replacements. Politicians intend to counter this demographic change in the short to medium term by taking in refugees, and are recording initial successes in this respect, because after all "one in three refugees has found work".[12] However, there is often a lack of adequate training and first of all a lack of language skills. In the 1980s, it was still possible to employ people in the company who could barely read and write or were not proficient in German, but with increasing digitalisation and specialisation, fewer and fewer simple jobs are available.[13] Other measures to counter

[9] Cf. Institute for Employment Research (2013), Federal Statistical Office (2008), Federal Employment Agency (2011).

[10] Cf. Rumpfhuber (2013).

[11] Cf. Mayer (2019).

[12] Cf. Bollmann and Kloepfer (2019).

[13] Cf. Militzer (2010), p. 74.

the impending shortage of skilled workers include creating incentives for couples to have offspring, e.g., by building residential property at affordable prices.

1.2 Places of Work

There is already a lot of research and publication on the requirements for the design of office and administrative buildings, but less beyond that:

> Scientific findings on objects other than office and administrative buildings hardly exist. In Germany, this is a young discipline that contains no standardized approach.[14]

Against the background of the fact that the various activities in the office have very different requirements, which have not yet been taken into account by appropriate spatial concepts, Gatterer calls for a specialisation of office equipment based on the following typology of knowledge workers[15]:

- Anchor (desk surfer): typically production managers, accountants, R&D managers who spend 90% of their time at their desk. They value a soothing atmosphere and comfort in the workplace.
- Connector (facilitator): typically planners, developers, lawyers who are 50% at their own desk and 50% within their own company. Their spectrum ranges from intensive meetings to concentration alone, which is why they need to be able to change the scenery quickly in order to work on a wide variety of topics.
- Gatherer: typically junior consultants, designers, marketing managers who also spend 50% of their time at their desk or its immediate surroundings, but also travel a lot outside the company. For them it is important to be able to communicate easily.
- Navigator: typically key account managers, trainers, senior consultants who use their office only 10% of their working time and often have no desk of their own. Their activity consists mainly of communication and the processing of information into knowledge.

Somewhat more finely structured is the differentiation of Jurecic et al. who identify the following seven types of work[16]:

- Thinker: Concentrated still worker who is also often on the move
- Hyper cross: Is everywhere and nowhere, highly communicative with colleagues, little at work, more in spontaneous meetings

[14] Essig (2020), p. 74.

[15] Cf. Gatterer (2009), p. 25 ff.

[16] See Schnell (2018) and Jurecic et al. (2018).

- Traveler: Always on the move, at home, in the office or on a business trip, when he is on site, he is in meeting rooms.
- Silent worker: Spends most of his working time at his desk and rarely has meetings.
- Communicator: travels a lot internally to discuss things with colleagues
- Caller: telephones a lot, switches frequently between projects and rarely works long on one activity
- Hands-on: Frequently switches between desk and lab or workshop

This classification can serve as a basis for the optimal office design, making sure that each type gets an office with the most suitable layout for him. A selection of office shapes is available for this purpose:

- Traditional single office
- Group offices for two, up to five or up to 20 people
- Open plan office
- Combined office (a mixed form of individual and open-plan office with communal zone)
- Multispacer offices with value-added offerings and sharing concept

Of these numerous office forms, the multispacer office is increasingly becoming the focus of interest because it serves the needs of very many types of work. This office structure offers flexible workplaces with desk sharing and a wide variety of meeting rooms, as well as retreats for concentration, recreation and breaks.[17]

In manufacturing, the majority of employees are still fixed to a fixed workstation. Of the types of work described above, the anchor is probably found most frequently in production, if one equates the desk surfer with the machine operator, because he is practically assigned to a single work point. Depending on the focal points of activity, three typical work locations emerge for a production employee involved in value creation, with the time shares in the activity depending on the position of the individual employee:

- Workplace
- Workshop
- Meeting

Workplaces in the manufacturing sector are structured very differently, whether in terms of location (e.g., natural light, high noise levels, etc.), occupancy, layout, and so on. Nevertheless, it is worth checking whether the structures found for the office workplace can also be transferred to industrial workplaces. For colleagues with different activity focuses at different locations, for example, consideration could be given to introducing Activity Based Working in a similar way to office planning or to taking it into account when planning a new building.[18]

[17] Cf. Steelcase (2013), p. 35 ff.

[18] Cf. Gatterer (2009).

The fourth industrial revolution, i.e., the path from the digitized world to the world of the Internet of Things and Services, is in full swing. Production will become increasingly intertwined with information technology. The goal is the interaction between man and machine[19]:

> The vision is the networked factory, where production data is available in real time and serves as the basis for further calculations. In order to fully exploit the possibility of digitalization, competencies must be built up and knowledge dispersed.[20]

Surveys show that the relevance of this topic is seen in many industries and is being addressed at high speed. The first applications are in the testing and implementation phase.

Automated activities with high routine shares reflect the developing digitalisation of processes. This progression contributes to the loss of pure production jobs, but in turn, new occupations will be created. This has led the German government to launch the "Industry 4.0" initiative, which aims to enable industry to be ready for the future of development and design.

The term Industry 4.0 refers to the fourth industrial revolution, at the end of which is the networking of all machines, products and processes in a "smart factory". This should serve to conserve resources, increase efficiency, increase the degree of automation for small batch sizes and increase quality through better process control.[21] Industrial production will be characterized by a strong individualization of products under the conditions of highly flexible production, the extensive integration of customers and business partners into business and value creation processes, and the linking of production and high-quality services.

Now, one might ask why the production workplace, of all places, should be endowed with a sense of well-being and the health of the employee should be emphasized, when there will be no more people working productively anyway, because this will be done by robots in the future:

> Work, however, is never done; at best it is taken over by automata. Freed from monotonous work routines, people are released into general unemployment. This utopia of leisure, however, does not begin after a machine revolution far in the distance, but has already begun with restructuring.[22]

This sentence is exaggerated, but in essence it is correct: Can robots perform any required activity in the short to medium term and displace humans?

---

[19] Cf. Walter (2018).

[20] Jurecic et al. (2018).

[21] Cf. Rauschnabel (2015).

[22] Rumpfhuber (2013), p. 75.

The ability of a robot to learn things and to develop constructively is already visible today. However, it should be borne in mind that the further development and improvement of the robot, as well as its manufacture, programming, maintenance and servicing, while taking less and less time, necessarily requires a human being. In the medium to long term, the professional use of a robot will undoubtedly come about, for example in industry, as is already the case in the automotive industry. The situation is different, however, when it comes to the development of household robots for all conceivable services, right up to measures in the care sector. Here, more is required of the machine than welding or painting. The use of artificial intelligence (AI) may therefore soon become the growth engine of the entire industry.[23]

1.3 Stress Factors and Well-being Factors

Before we shed light on the effectiveness of well-being factors, let us take a look at the stressors they face.

In production, research into well-being and health is largely concerned only with the office and administrative departments, but for employees in the actual production process only two key figures are essentially known, turnover at around 10–15% and the sickness rate at around 2–4.5%. As soon as the efficiency of the production process decreases, the production time will increase by the targeted production quantity, which may result in these ratios deteriorating. A reduction of the efficiency of about 5% makes a sum of 2.450.500.000 € with an average salary of about 3100 € per employee and 13 million employees.

When optimizing work processes, all too often only financial aspects are in the foreground and people are often neglected. This manifests itself in the form of higher pressure to perform, triggered by greater responsibility, and in the resulting stress, which in the long term can lead to demotivation and also to illness. A survey of 1660 employees in March 2016 revealed the following stress factors, with multiple responses possible[24]:

- Constant deadline pressure
- Poor working atmosphere
- Emotional stress
- Overtime
- Permanent readiness
- High pressure to succeed
- No breaks or too short
- Monotonous tasks

[23] Cf. Weiguny (2017).

[24] Cf. pronova BKK (2018), p. 24.

- Poor workplace equipment
- High physical stress
- Mobbing
- Shift work

These stressors affect physical and mental well-being in a variety of ways[25]:

- Back pain
- Persistent fatigue and exhaustion
- Inner tension
- Brooding
- Listlessness
- Sleep disorders
- Irritability
- Headache
- Concentration problems
- Diminishing sexual desire
- Withdrawal from circle of friends/acquaintances
- Stomach/digestion problems
- Self-doubt

To reduce stress, the following measures are recommended:

- Sports
- Relaxation techniques such as yoga, autogenic training or progressive muscle relaxation according to Jacobson
- Small periods of relaxation throughout the day, e.g., a walk during the lunch break or a relaxing bath in the evening.
- Fixed and regulated daily routine
- Healthy diet with an optimal mix of nutrients

However, it is equally important to prevent this from happening in the first place. The employer also has a duty here, because the establishment of health-promoting conditions and the development of healthy behavior in conjunction with the design of a health-promoting working environment is the basis for the protection of employees.

The Workspace Futures team at Steelcase, a leading developer of workplace environments, was able to identify six dimensions that influence well-being and that a modern workplace should promote[26]:

- Optimism: Creativity and innovation require optimistic people who keep the big picture in mind, pursue ideas, are open to change and take the risk of facing difficult tasks.

[25] Cf. pronova BKK (2018), p. 32.

[26] Cf. Steelcase (2013), p. 21 ff.

- Mindfulness: The fast pace, volatility and rapid technological advancement of today's working world cause cognitive overload and stress. With mindfulness, a balance can be found with the actual here and now, allowing one to be attentive, focused, and fully engaged and passionate about one thing.
- Authenticity: The feeling of being oneself and being able to express oneself freely creates a sense of well-being.
- Belonging: Social relationships, friendships and positive interactions at work create the basis for feeling connected to a company, its brand and its goals.
- Meaningfulness: Pursuing a meaningful goal is a prerequisite for motivation and a positive attitude towards work. People need meaningfulness in order to understand that their work is not superfluous.
- Vitality: The permanent inactivity of muscles, as happens when sitting for too long, impairs attention, whereas changing posture frequently stimulates the mind. Therefore, sufficient movement at work is essential to be physically and mentally fit.

For this reason, it is necessary to work on the promising well-being factors which, on the one hand, bring benefits to both employees and the company and, on the other hand, ensure a balance between the world of work and private life (work-life balance).

Conclusion

The continuous change from an industrial to a knowledge society puts the typification of jobs more and more in focus and will become even more important. In order for employees to be able to keep up with this change with all its new demands, a preoccupation with well-being factors is imperative.

References

Bollmann, R., Kloepfer, I. (2019), Jeder dritte Flüchtling hat Arbeit gefunden, in: Frankfurter Allgemeine Zeitung, Wirtschaft, 11.08.2019.

Bundesagentur für Arbeit (2011), Arbeitsmarkt 2011, Bundesagentur für Arbeit, Nürnberg.

Essig U. (2020), Ein Heim für die Seele, Apotheken Umschau, 15. Januar 2020, S. 74–76.

Gatterer, H. (2009), Räume der Arbeit, Trendreport zu Büro und Arbeitswelten, Grasl Druck & Neue Medien, Wien.

Institut für Arbeitsmarkt und Berufsforschung (IAB): Daten zur kurzfristigen Entwicklung von Wirtschaft und Arbeitsmarkt, 04/2013.

Jurecic, M.; Rief, S.; Stolze, D. (2018), Office Analytics – Erfolgsfaktoren für die Gestaltung einer typbasierten Arbeitswelt, Fraunhofer Verlag, Stuttgart.

Mayer, T. V., (2019), Die Demographiewende ist machbar, in: Frankfurter Allgemeine Zeitung, 08.07.2019.

Militzer, H. (2010), Der Novocain-Betrieb, Geschichte und Geschichten, Hoechst AG, Aventis, Frankfurt am Main.

pronova BKK (2018), Betriebliches Gesundheitsmanagement 2018, Ergebnisse der Arbeitnehmerbefragung | Februar 2018, https://www.pronovabkk.de/media/downloads/presse_studien/studie_bgm_2018/pronovaBKK_BGM_Studie2018.pdf [Zugriff am 08.06.2020].

Rauschnabel, J. (2015), ISPE Fachgespräch, Leipzig, 24. Februar 2015.
Rumpfhuber, A. (2013), Architektur immaterieller Arbeit, Tuna+Kant, Wien.
Schnell, S. (2018), Wie verschiedene Mitarbeitertypen und Arbeitsstile die Büroplanung beeinflussen, https://business-user.de/arbeitswelt/wie-arbeitstypen-und-stile-die-bueroplanung-beeinflussen/ [Zugriff am 25.05.2020].
Seiferlein, W.; Kohlert, C. (2018), Die vernetzten gesundheitsrelevanten Faktoren für Bürogebäude, Springer Vieweg, Wiesbaden.
Seiferlein, W.; Woyczyk, R. (2017), Projekterfolg, die vernetzten Faktoren von Investitionsprojekten, Fraunhofer Verlag, Stuttgart.
Statistisches Bundesamt (2008), Klassifikation der Wirtschaftszweige, Ausgabe 2008 (WZ 2008), Statistisches Bundesamt, Wiesbaden.
Steelcase (2013), Wohlbefinden – ein Thema, das nur Gewinner kennt, 360° Magazin Nr. 8, https://www.steelcase.com/content/uploads/sites/2/2018/08/360N8DE.pdf [Zugriff am 08.06.2020].
Steelcase (2016), Mitarbeiterengagement und Arbeitsplätze in aller Welt, 360° Steelcase Global Report, https://cdn2.hubspot.net/hubfs/1822507/2016-WPR/DE/SteelcaseGR_DE.pdf [Zugriff am 08.06.2020].
Walter, G. (2018), Trending Topics – Industrie 4.0, https://www.faz.net/asv/trending-topics/trending-topics-industrie-4-0-15916467.html [Zugriff am 04.06.2020].
Weiguny, B. (2017), Erfolgsformel: Arbeiten 4.0 und Führung 4.0, in: Frankfurter Allgemeine Zeitung, 23.04.2017.

State of Research: Measures for Well-being in Different Areas

2

Preliminary Remarks
In the following, initial factors regarding well-being are identified and selective measures already implemented are presented in order to show the status of this issue in the production and industry.

2.1 Measures on the Building Envelope: Implementation of a Colour Concept

Friedrich Ernst von Garnier made a name for himself with colors in industry in the 1970s and 1980s. Initially, he devoted himself to coloring the outer skin of production building facilities and designed the colors of more than 70 buildings and facilities at Industriepark Höchst.

Von Garnier's work is regarded as proof that architecture, buildings and production facilities need not be dominated by dreary grey. His doctrine of "organic colourfulness" is based on the necessity of multi-tonal colourfulness for the maintenance of well-being and thus ultimately health.[1] Colourfulness is thus not superfluous decoration, but thoughtfully designed an inseparable part of the human soul.[2] Thus, Garnier was commissioned to design the colours of three Hoechst AG plants.[3] In the Novocain plant, the walls were to be painted corn-yellow and the ironwork lime-green, while sky-blue was intended for the

[1] Cf. Garnier (1997).

[2] Cf. Militzer (2010).

[3] The Novocain and Dolantin operations and the North Technical Center.

W. Seiferlein, *Well-Being Factors for Different Industries*,
https://doi.org/10.1007/978-3-658-34997-4_2

Fig. 2.1 The Novocain factory building in its original colour scheme

engines of the stirred tanks. The plant manager Hans Militzer describes the effect as follows:

> When the project was implemented, the design seemed to take some getting used to at first glance. At the Höchst plant, the Novocain plant was referred to as a candy factory. However, if you let the colors sink in a bit, you could relate to what had prompted von Garnier's choice. He wanted to bring positive associations of nature into the mechanized working world with its boilers, equipment and pipelines in order to make the working world more human. The corn yellow makes you think of a ripe cornfield in summer, when the first leaves are sprouting, and you can see the sky between the leaves.[4]

The key to success was therefore not the selection of a single colour, but its implementation by means of a colour concept (Figs. 2.1 and 2.2).[5]

Conclusion As a success factor for the effect on the well-being, the choice of colours with a suitable concept is to be mentioned quite clearly.

[4] Militzer (2010), p. 79 f.

[5] Cf. Trautwein (2018), p. 131.

Fig. 2.2 The Novocain factory hall with friendly colour scheme

2.2 Indoor Measures: Nature and Art

Stress has a negative impact on health, but contact with the natural environment can reduce the effects of stress. Furthermore, it can regulate blood pressure and heart rate and improve people's positive attitude and overall happiness.[6] A look at the implications for clinical patient outcomes shows similar findings. Patients with the view of trees and/or water were significantly less anxious than other patients during the postoperative period. They suffered aless severe pain or switched from strong to moderate pain medications more quickly than other groups.[7] This awareness can also be applied to industrial production turing sites.

The planners at Merckle, for example, used their location advantage to place manufacturing in a heavily landscaped environment. The advantage of the window view is the visual connection to nature, which can improve employees' mental engagement and attention. As Fig. 2.3 shows, the operator's gaze inevitably goes to the greenery. Thus, even when waiting, there is an opportunity to create well-being.

Instead of a window view, the view of an object, i.e. pictures or photos, is also possible. The objects should have a generally relaxing effect with regard to the motifs and be

[6]Cf. Aldwin (2007).

[7]Cf. Ulrich (1984).

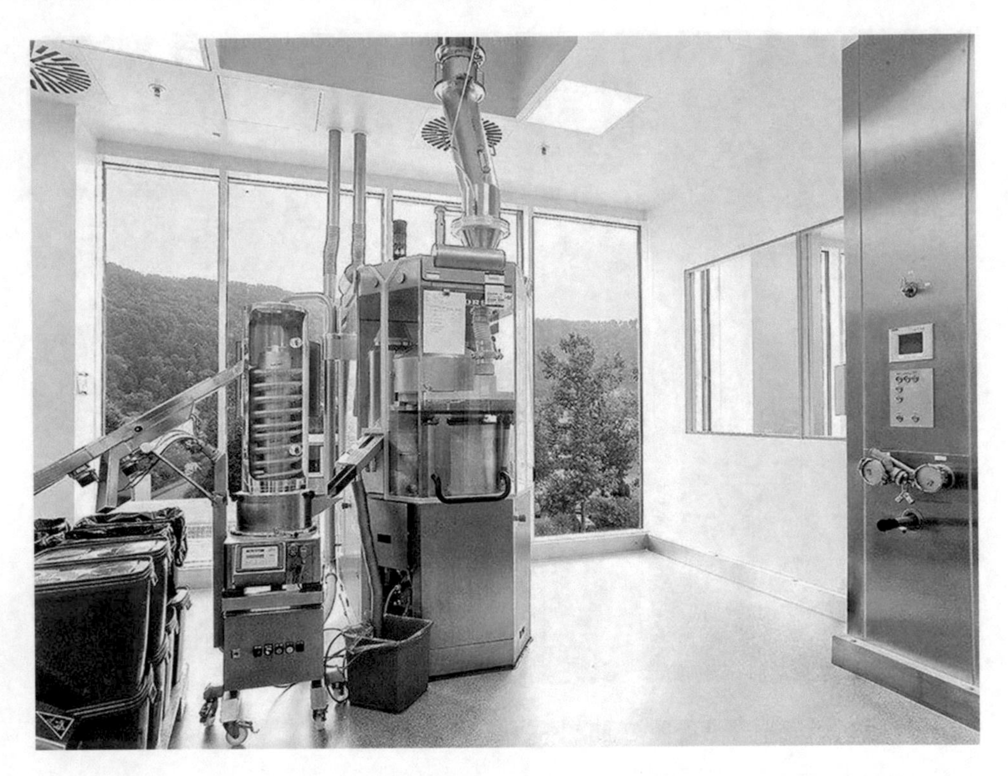

Fig. 2.3 If the process permits, the operator can enjoy the view of the greenery (Merckle)

mounted on the selected wall in an appropriately dimensioned manner. Another option is an “illuminated picture”, which as a background motif can convey a good feeling to the employees working there as well as to visitors, auditors and guests.[8]Any production room or social room can be used as a location (Figs. 2.4 and 2.5).

Salingaros evaluated various studies on stress-reducing environments, which had found out that the human organism reacts particularly positively to fractal manifestations, i.e. to sequences of self-similar structures. According to Salingaros, the greatest effect, with a stress reduction of 44%, is shown by a medium value of the fractal dimension, as also occurs in snowflakes, coastlines, plants or vein systems.[9]

Kellert et al. also emphasize the positive effects of nature on human health.[10] They combined findings from neuroscience and biology in their theory of biophilic design in order to make them practically usable. Architecture and interior design inspired by nature makes it possible to reduce physical stress and turn workplaces into liveable, productive and healthy places. This includes biophilically designed art, which creates a contrast to the

[8]Cf. Sect. 5.5.

[9]Cf. Salingaros (2012).

[10]Cf. Kellert et al. 2013.

Fig. 2.4 Photo wallpaper at Seegmüller Darmstadt, "Holiday scene in Tuscany"

Fig. 2.5 Photo wallpaper at Seegmüller Darmstadt, " **Blooming Erica** fields in northern Germany"

Fig. 2.6 Insertion of art in the clean room areas at Merckle

usual work environment and encourages people to pause and see the world from a different perspective. The goal is to use art to briefly relax the eye, the body and the mind.

For this reason, the Merckle company decided to display art from the nearby environment for the employees in the production of products manufactured under cleanroom conditions. In order to comply with the handling according to the clean room regulations, new GMP-compliant display cases were constructed and installed. The art objects, mainly paintings, were placed in glass cases that meet the usual conditions of cleanroom equipment (Fig. 2.6).

The creative proposal by Carpus & Partner to install floor-to-ceiling windows was also discussed, but then rejected for rational reasons of occupational health and safety (Fig. 2.7).

The establishment of this company "gallery" also opens up other possibilities. For example, a company competition could be launched with pictures or photos of the employees. This increases recognition, the "we-feeling", etc. With these objects, the form of a temporary exhibition would also be conceivable.[11]

[11] Cf. Sect. 5.3.2.

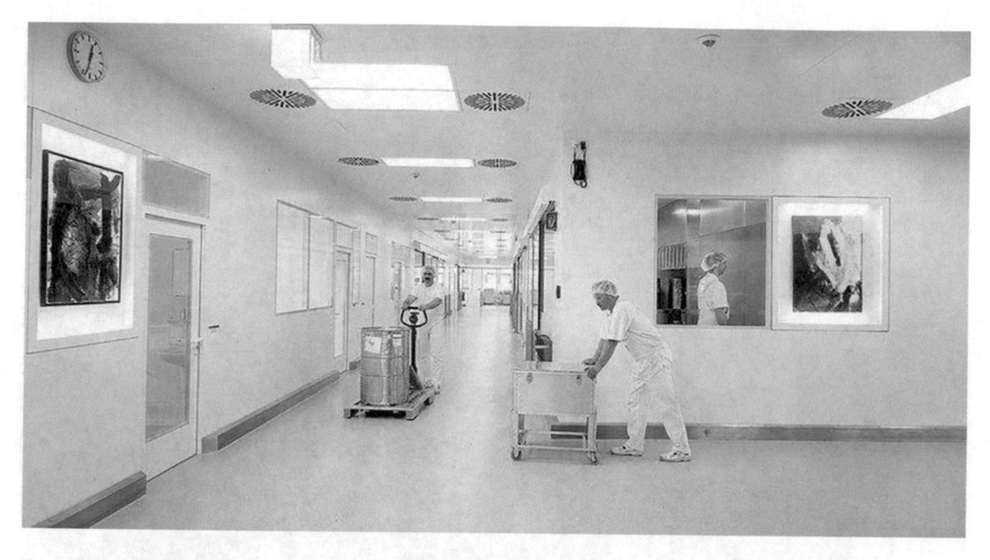

Fig. 2.7 Insertion of art in the clean room areas at Merckle

Conclusion In this case, the investment costs for the art and its "accommodation" have to be spent, which, however, in comparison to the then expected decrease in sick leave and reduced fluctuation, can be classified as a calculable project for the true appreciation of the employees.

2.3 Measures Relating to Architecture and Social Spaces

Public spaces are not only available for communication but also for other social functions. Many companies have a public space in order to be open to external visitors or to introduce the plant to internal colleagues and their children, for example, who do not yet have access. The company Novartis stands out in this respect as a general measure with a master plan and broad-based concepts. Merck in Darmstadt is acting in the same direction.

Expressive architecture defines the guiding principle of the Novartis master plan.[12] But Novartis is not only demonstrating a good hand in allocation and exterior architecture, but also in terms of interior design and with art. At the Novartis campus in Basel, an open approach to art is desired,[13] with the same goal as at Merckle of improving the sense of well-being of employees and promoting "communicative work and intensive exchange of information"[14] (Figs. 2.8 and 2.9).

[12]Cf. Novartis (2009), p. 106.

[13]Cf. Novartis (2009), p. 104.

[14]Cf. Novartis (2009), p. 106.

Fig. 2.8 Seating by the daylighting atrium

Fig. 2.9 Atrium building

From this, the scope of this topic can be derived and the importance of well-being can be shown. However, these measures are mostly selective and not designed as a concept. This is not the case with a current project, namely the establishment of a well-being university. The seven universities in Cologne (Germany), Birmingham (UK), Florence (Italy), Leiden (Netherlands), Linnaeus (Sweden), Nantes (France) and Semmelweis (Hungary) founded the "European University for Well-Being"(EUniWell). The aim of the partnership is to foster an environment "in which both Europeans and their global neighbours can be highly educated, socially engaged, healthy, inclusive and diverse". By the concept of "well-being", the universities mean quality of life and well-being,[15] i.e. precisely the two parameters that are at stake when talking about well-being factors.

Conclusion

For the purpose of beautification and under psychological aspects, mainly singular measures per application site were also realized at Novartis. Of course, these individual measures also improve well-being. However, with the knowledge of other factors, it is not difficult to put together well-being concepts according to requirements. These individual measures really call for the development of a concept or master plan that shows the possible measures from the various sectors or areas and the respective project phase and that also includes the office buildings.

References

Aldwin, C. M. (2007), Stress, coping, and development – An integrative perspective, Guilford, New York.

Blazekovic, J. V. (2020), Unis gründen Netzwerk für das Wohlleben, Sozial, Gesund und Inklusiv, in: Frankfurter Allgemeine Zeitung, 08.02.2020,

Garnier, F.-E. V. (1997), Meine farbigere Welt – Ein ganz unsachliches Sachbuch, Verlag Matthias Ess, Bad Kreuzbach.

Kellert, S. R., Heerwagen, J., Mador, M. (2013), Biophilic Design, The Theory, Science and Practice of Bringing Buildings to Life, John Wiley, New York, S. 3–20.

Militzer, H. (2010), Der Novocain-Betrieb, Geschichte und Geschichten, Hoechst AG, Aventis, Frankfurt am Main.

Novartis (2009), Novartis Campus, Hatje Cantz Verlag, Ostfildern.

Salingaros, N. A. (2012), Fractal Art and Architecture Reduce Physiological Stress, in: Journal of Biourbanism 2/2012, S. 11–28.

Trautwein, K. (2018), Farbkonzepte für Arbeitsplätze, in: Seiferlein, W., Kohlert, C., Die vernetzten gesundheitsrelevanten Faktoren für Bürogebäude, Springer Vieweg, Wiesbaden, S. 102–114.

Ulrich, R. S. (1984), View through a window may influence recovery from surgery, in: Science 224(647), S. 420–421.

[15] Cf. Blazekovic (2020).

3 A Look into Practice: Interviews and Literature Research

Preliminary Remarks
The aim of this book is to evaluate known and still unknown well-being factors in different industries and sectors The purpose is to generate a broadly based tool. This tool will be called "Target Matrix", for the most part, interviews or discussion rounds have been conducted with the relevant industry representatives. Where no interview partners could be found, the findings came from literature research and on-site visits. Well-being factors were identified in the following areas:

- Office/laboratory construction
- Hotel business
- Shopping centre
- Cruise ship
- Idea train
- Service provider
- Banking sector
- Airport
- Urban planning
- Garden (facilities)

It can initially be assumed that the well-being factors identified are also applicable to people in other sectors and industries. However, this is an unconfirmed assumption. However, since the well-being factors are essentially shaped by people, the respective stimuli are likely to have an effect on every person, although the nature of the activities differs. For industrial production, existing factors would have to be tested for their applicability and, if necessary, new factors would have to be added.

W. Seiferlein, *Well-Being Factors for Different Industries*,
https://doi.org/10.1007/978-3-658-34997-4_3

The interviews were based on a guide of questions covering the main areas of interest in the research:

- What is a well-being factor for you?
- How do you determine the well-being factors?
- Are there statistics on the effectiveness of the factors among users?
- Did you have role models to fall-back on?
- How do you consider well-being factors in planning?
- Was the participation in the preliminary planning designed for one or more employees?
- Is the planning team interdisciplinary?
- Are structured solution-finding processes such as mind maps, brainstorming or programming also used?
- Do you involve specialists to implement the well-being factors?
- What results do you expect (satisfied customers and employees, lower sick leave, etc.)?

In the following recordings of the interviews, the questions (Q) were asked by the author, the answers (A) were given by the respective expert. At the end of each chapter, the most important points are summarized and the newly added well-being factors are listed.

3.1 Office and Laboratory Building

The scientific investigation for well-being factors is most advanced in office and laboratory construction. The following is a list of the relevant factors that emerged from the investigation in this area.[1]

Demand-oriented Building Services Engineering, Layout of Office Areas
The rooms should have a calming effect through materials, surfaces, colours, light and view. Sufficient space, safe accessibility of the premises and work equipment, opportunities to alternate between standing and sitting, enough storage space for personal belongings and sustainable materials promote well-being.

Adequate Office Equipment
The office furniture for the workplace, the furniture for the different areas of collaboration and exchange, concentration and relaxation as well as the IT equipment should have an appealing design, be intuitive to use and follow ergonomic principles. This supports different sizes, needs and preferences and enables movement and communication at the workplace.

[1] Cf. Seiferlein and Kohlert (2018).

Work-life Balance

In order to achieve a balance between the personal interests and needs of employees and the strategic goals of the company, both sides must pull together and jointly develop a vision of the corporate culture based on mutual appreciation, open communication and leadership with trust. This can be done through professional change management and the development of an Employers & Company Alignment.

Individual Perceptions

The factors of air (**L**uft), noise (**L**ärm), (**L**licht) and body (Laib), which can be summarised as the 4 L® in German Language, can be felt directly in one's own body and therefore have a direct effect on well-being[2]:

- If the **air** is too cool or too warm, too dry or too humid, we feel uncomfortable, even a slight draft is perceived as unpleasant.
- **Noise** disturbs concentration, whereas a moderate background noise or background music can promote relaxation.
- The **light** at the workplace should not dazzle, should be neither too bright nor too dark and should be individually switchable. It makes sense to integrate the lighting into a colour concept.
- The **body** factor includes all things related to eating and drinking.

Colour Concept

Colours in light make the objective world visible to us. Matching the colours in a room to the lighting concept is therefore an important prerequisite for the design of pleasant working environments. People see bright things first, which is why the places and surfaces that should attract the most attention should be designed brightly.

Medical Aspects

Hygiene, cleanliness, occupational health and safety and health-promoting measures contribute to the well-being of all and can help to reduce sick leave. These include regular cleaning of the office and work equipment, creating opportunities for movement in the workplace or supporting sporting activities. The latest trend is pets in the working environment.

Conclusion

In office and laboratory construction, well-being factors have already been very well researched, so they can serve as a basis for further identification of factors from other sectors and industries.

[2] Cf. Seiferlein and Kohlert (2018), p. 31 ff.

Well-being Factors from the Field of Office and Laboratory Construction

- Architecture: functional room layout
- Ergonomics: adequate furniture, office furniture, IT equipment
- Work-life balance: mutual agreement, appreciation, Employer & Company Alignment,
- Perception: air, light, noise/music, body (Change management 4 L®), colour concept
- Nature: animals
- Health: hygiene/allergies, quality, exercise

3.2 Hotel Business

In a study conducted by the Fraunhofer Institute, the "Future Hotel Guest Survey", approximately 3380 hotel guests in the DACH market were surveyed in order to obtain the functional user-related requirements of a hotel such as equipment features, individual hotel room selection, etc..[3] This served to gain insight into the needs of different target groups and as a basis for developing meaningful solutions for the industry in line with requirements. Numerous hotel operators of different quality levels and locations, who were not available for an interview due to a lack of resources, confirmed to the author of this book that they work with well-being factors. However, these used to be named differently and were not systematically processed into measures.

The hotel selected for the interview belongs to a hotel chain that was founded in the early 1980s and initially focused on the main holiday destinations in Spain. The company currently owns over 60 hotels in 20 destinations with a total of more than 17,000 beds, making it one of the ten largest hotel companies in Spain. Most of the hotel properties are owned by the hotel chain. The company places emphasis on offering customer-oriented services that are constantly adapted to the current needs of its guests, in order to guarantee them a quality stay. The continuous implementation of high quality standards in daily activities is laid down in a certified quality management system, which has already received several awards, and integrates prevention, safety and sustainability. The managers, directors and employees of the hotels as well as all customers, partners and suppliers are required to follow this code of ethics.

The interview was conducted with the manager of the hotel who wishes to remain anonymous (Fig. 3.1).

(Q): Apparently the hotel is in good shape not only externally but also in terms of process, what are the reasons for this?

(A): The management recognized at an early stage that systematic quality improvement is of paramount importance. As proof of our activities, we were awarded the internationally recognised ISO 9001:2015 certification, among others

[3] Cf. Borkmann et al. (2013).

Fig. 3.1 General view of the plant

(Q): Are well-being factors involved in the quality activities?

(A): Yes, quite hotel-specific factors come to mind, such as size of the balcony, length x width of the bed, quality of the mattress, etc. (Fig. 3.2 and 3.3).

(Q): How do you define appropriate well-being factors?

(A): Such factors are described in a standard that is binding for all locations. This specifically includes, in addition to overriding well-being factors such as well-trained, competent and friendly staff, the planning of garden areas with many palm trees, well-maintained water features such as fountains, waterfalls, etc. and, of course, with many green plants throughout the facility area. Other factors include scents, background music and water features. Scents and smells have a greater influence on us than we think. Findings from neuroscience confirm that scents directly influence the limbic system—and thus feelings and memory. "Scents touch and discover memories."[4]In addition, fragrances are used to increase clientele. That's why we already use scents in all our public spaces. Background music

[4]Cf. Tran (2019), p. 55.

Fig. 3.2 Light

creates a certain atmosphere; in the restaurant, for example, the music invites people to linger or even to leave the restaurant (Fig. 3.4).

(Q): Hotels with many wellbeing factors are probably priceless, can you confirm that?
(A): With greater demand and increasing quantities, such factors become cheaper over time. This effect is also often observed in the economy.

Another argument is that one has to select from the identified factors a choice that is woven into a fine, sophisticated concept. It will then be easier to respond to the individual wishes of the guest if the visual proportions, acoustics (music), temperature, light and smell are perfectly matched (Fig. 3.5).

(Q): What is the process of project planning?
(A): The simplest projects are internal projects, which are usually financed with budget money. Relevant resources and functions are brought in for approval, depending on the topic and needs. The usual composition of the project team, depending on the subject area, is the technical service, operational guest support, management, etc. The project team is deliberately staffed with interdisciplinary members. If a project

Fig. 3.3 Air

involves higher levels of investment, the global department is involved with an architect and civil engineer (Fig. 3.6).

Conclusion

Remarkable is the consistent implementation of the quality guidelines and the integration of a structured colour, acoustic and communication concept. As a knowledge gain, a vital result has emerged. In addition, the question can be pursued which influencing factors have an effect on the recovery of guests.

Well-being Factors from the Hotel Sector

Newly added factors are marked in bold.

- Nature: **plants, water**
- Perception: light, air, noise/music **(acoustic concept)**, body (4 L®), colour concept, **scents**
- Health: hygiene, exercise, **place of rest**
- Ergonomics: adequate furniture

Fig. 3.4 Place of silence

- **Communication concept** (guests, employees, etc.)
- Work-life balance: appreciation

3.3 Shopping Centre

No interview partner could be found for this area, so this chapter was based on the specialist literature on the subject.[5]

According to Lange, the following points should be considered as success factors for shopping centres[6]:

1. Location and accessibility

[5]Cf. in particular Lange (2009) and besides Bastian (1999), Franke (2007), Gruen and Smith (1960).
[6]Cf. Lange (2009), p. 59 ff.

Fig. 3.5 Music—direct or as background music

Catchment area, transport links, competition, target groups

2. Architecture and interior design

   Urban and architectural form, internal development, climate and lighting, design and decoration, parking spaces

3. Anchors and satellites

   Importance of anchors and satellites, economics of anchors and satellites, rent levels

4. Sector mix and coupling potential

   Clustering of product-related stores, coupling potential of stores and branches, optimal arrangement of stores and branches, time management.

For the identification of the well-being factors, the aspects of architecture and interior design as well as location and accessibility are of particular importance.

Lighting

Pleasant lighting increases the feeling of well-being, especially if a warm light and indirect lighting is used, as "blue" light and direct irradiation have an aggressive and uninviting effect.

Fig. 3.6 An opportunity for sport and play

Climate

The climate, i.e. the optimal ratio of temperature, humidity and air speed, is a quality criterion for a shopping center and influences the well-being of the people who spend time in it. In order to prove this empirically, a study in the USA determined the distance and the time that a healthy person still wants to walk under different climatic conditions. The result illustrates the significant influence of the climate on the willingness of customers to walk and the time they spend there, because they were willing to walk four times longer and further in an air-conditioned shopping mall than in an uncovered shopping street.

Interior Design and Decoration

For higher-end shopping centres, as with office worlds, the points of interior design and decoration will be important parameters for increasing their appeal to customers. This requires a concept that appeals to different target groups and follows a corporate identity. The entire furnishing from the design of the floor, handrails, signage and shop fronts to special installations (e.g. fountains, lights, mosaics) must be coordinated with this.

There is a consensus among experts that the atmosphere as well as the experience factor is significant for sophisticated shopping centers, because many customers do not simply buy essential goods, but practice shopping as a leisure activity. Through a coherent overall concept of architecture, interior design and decoration, each shopping center gains its very own, coherent character and communicates a certain image that is intended to appeal to the target groups sought, increase their length of stay and maximize the probability of targeted and spontaneous purchases.

Pitches

The point of providing parking facilities is valid today and must be taken into account as a success factor for the well-being of employees and customers, although new mobility offers can also be expected in the medium term. In the future, mobility and status thinking will change. For example, the trend is away from owning a car towards public transport, which also includes car sharing, but e-mobility and bicycles will also gain in importance. A good example is the Scandinavian countries, which already implemented such systems 20 years ago.

Car parking facilities are necessary for shopping centres outside the cities, because they live from car traffic, but also for shopping centres in the city centre, because the parking space supply is usually limited there.

The importance of parking facilities was shown, among other things, by a comparison of success factors in the retail sector, in which parking spaces were named as the most important criterion. In second place was the criterion of accessibility, and only in third place were the factors of sector mix, variety of offers, etc. named.

Well-being Factors from the Shopping Centre Area

Newly added factors are marked in bold.

- Architecture: Functional space layout, one **parking** space for each vehicle of the employees and customers
- Perception: air, light, noise (4 L®)

3.4 Idea Train

> The mobility behavior of commuters in Germany is changing rapidly. Passengers expect comfort and individuality and want to use their time wisely. [...] Vehicles must be suitable for mass transport and highly reliable, but at the same time comfortable and individual.[7]

Many of the passengers' wishes for services on the train and for the design of a feel-good atmosphere in the passenger compartment were incorporated into the development of the Ideenzug. The development of a feel-good atmosphere plays a special role in the Ideenzug, because this is the only way to achieve the project's goal of making passengers feel safe and comfortable in the trains.

[7] Deutsche Bahn (2017), p. 1.

The interview was conducted with Mr. Carsten Hutzler, one of the project managers of the Idea Train of Deutsche Bahn.

(Q): How was the project born?

(A): The kickoff for the project took place with the renowned Munich-based design agency Neomind. Together with Deutsche Bahn's innovation lab d.lab, the DB-Regio subsidiary Südostbayernbahn (SOB) and with the Bavarian Railway Company, creative and comfortable ideas and themed worlds were designed to make local rail travel even better in the future. The first step was to research current trends and conduct customer surveys. The results were compiled and evaluated. Workshops were held at regular intervals by the DB project team together with customers, in which the large number of different ideas were refined and thought through further. The target groups were also defined together.

(Q): How were the taFrget groups defined?

(A): The focus in designing the product and service worlds of the Ideenzug is on groups of people such as commuters, families, students and senior citizens. Of course, the customer advisory board, which represents the views of the customers, was also brought on board or into the train.

(A): Was there a role model for the integration of well-being factors?

(A): No, there was and is no role model of this kind. There are creative thought-provoking impulses from office construction, above all from the furniture industry, which could be adapted and adopted. What is important, however, is that one distinguishes oneself even more sharply from other modes of transport (Fig. 3.7).

(Q): What is the function of the well-being factors?

(A): In our project, well-being is described as a feel-good atmosphere. This description better captures the desired emotional level.

Rail customers wanted, for example, a sports studio, a single cabin for relaxed working, a lounge for group travel or, especially for families, a children's paradise. A total of 22 themed areas were developed with the following content:

- Catering and snack area
- Family area with separate play corner
- MyCabine for a retreat into tranquillity or privacy
- Relaxing with armchairs for (almost) lying down
- Comfort seats e.g. for working
- Lounge area for several people or groups

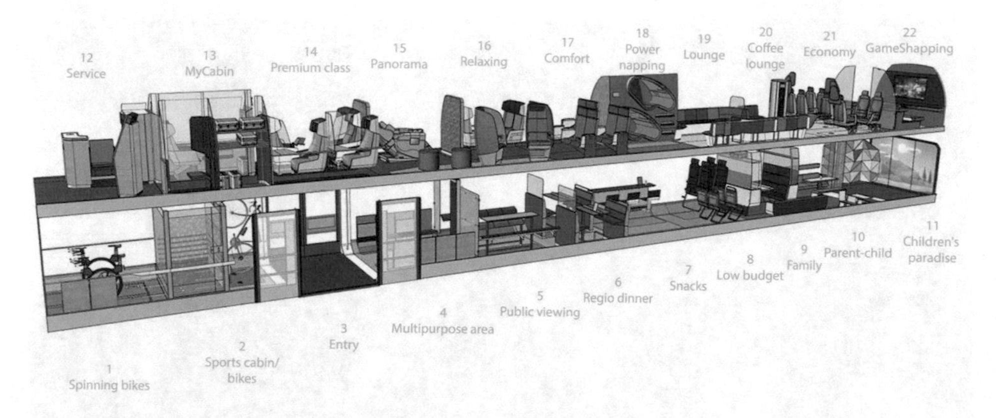

Fig. 3.7 Interior of the idea train

- Spinning bikes for sports activities
- Public viewing, e.g. to watch football with other travellers or to get information (Fig. 3.8).

(Q): Were any specific methods used to gather user requirements?

(A): Through brainstorming and creative workshops, all conceivable uses, functions and products for train interiors were collected, assigned to specific customer groups (e.g. commuters) and compared with their needs (Figs. 3.9, 3.10, 3.11 and 3.12).

Design Improvements Worth Mentioning

- **"Stand-up seat"**
  Due to its design, less surface area is used when standing/sitting
- **"Change seat"**
  By swivelling the backrest, for example, a four-seater (Vis-a-Vis) can be created on the left and a two-seater on the right.
- **"Power napping module"**
  It is very "simple" in design and execution, which looks functional and comfortable. Only the boarding needs to be more comfortable.

Well-being Factors from the Idea Train Area

Newly added factors are marked in bold.

- Ergonomics: adequate furniture
- Health: place of rest, **power napping, sports** at work
- Communication: **informal communication** (public viewing), communication concept

Fig. 3.8 Power napping area integrated in the interior

Fig. 3.9 The comfort area offers comfortable seats

Fig. 3.10 "Stand-up seat"

3.5 Cruise Ships

General Information about Meyer Werft

The Papenburg-based Meyer Werft was founded in 1795 and is now in its seventh generation of family ownership. Today, the company is best known for building large, modern and sophisticated cruise ships, which it started in 1985 with the construction of the Homeric—almost 200 years after it was founded.

The planning and construction of a passenger and cruise ship is a technically and logistically extremely demanding major project. Meyer Werft sees shipbuilding as a modern industry that combines proven knowledge and new technologies in one product, while offering technical and operational safety. The cruise ships are thus platforms for forward-looking tourism concepts. Established shipping companies from all over the world

Fig. 3.11 "Interchangeable seat"

commission new ocean liners from Meyer Werft.[8]In Papenburg, in Rostock at the Neptun shipyard and in Turku at the Meyer Turku shipyard, innovative cruise ships, ferries and river cruise ships are built in close cooperation. With the appropriate know-how in technology and engineering, luxury liners are built that meet the very highest standards. Computer-aided technologies for design, planning, construction and manufacturing are now integrated into all work processes.

The interview was conducted with Mr. Lars Kruse, Head of the Sales and Design Department (Fig. 3.13).

Identification of Well-being Factors

Instead of well-being factors, in shipbuilding one often speaks of comfort or ambience (ambient control). At TUI Cruises, the word "feel-good factor" has become an established term and ultimately means to fill passengers with fun and enjoyment.

[8]Cf. Meyer Werft (2020).

Fig. 3.12 A sound-absorbing armchair spatially adapted for the train of ideas

With the creation of user requirements, the most important and formal planning requirements are fulfilled. In order to maximise knowledge of user requirements, an experienced and future user (Passenger Consultant) is often involved in this planning step.

In the early phase of planning, all information is collected and integrated into the process. The planning team decides together on the parameters for comfort or ambience, i.e. the well-being factors.

Documents developed in a team promote a sense of togetherness. The composition of the team should be broad and interdisciplinary. An owner's representative from the shipyard is available to the team. In addition, an architect from the client's side is present in the team.

The Different Well-being Factors

Compared to other industries and sectors, there is an astonishingly high number of well-being factors in cruise shipping, which can be roughly divided into ship-specific and general factors. Ship-specific factors usually have their origin in technical conditions,

Fig. 3.13 The ships of the Meyer Werft shipyard

such as unwanted odours or vibrations, the origin of which is often unknown. The general factors are more interesting and more in focus at this point, because they can also be applied to other industries and areas. Almost all factors have a general character.

The safety aspect is very important, and here too the factors can be divided into ship-related and general ones.

The well-being factor of intellectual activity is a new factor. On many cruise ships there is a library with "real books". This place exudes coziness and invites you to browse the book selection list. In addition to books, board games such as Monopoly, chess or Scrabble can also be borrowed.

Well-being Factors from the Cruise Ship Sector

Newly added factors are marked in bold.

- Health: hygiene, **safety** (e.g. fire protection), sports and wellness offers (wide range of entertainment, Kids Club), **mental activity**
- Architecture: functional room layout (nice cabins = hotel rooms)
- Perception: noise, body (good, healthy food)
- Nature: **sea**

3.6 Service Provider

On the subject of service providers, two companies were interviewed: Sparkasse Dieburg on the one hand and Fraport at Frankfurt Airport on the other.

3.6.1 Savings Bank Branch

Sparkasse Dieburg opened a "feel-good branch" in June 2019, realising the idea of a branch of the future as a place that facilitates encounters and is familiar to customers.

The interview was conducted with Mr. Florijan Blazevic, Center Manager of the new branch of Sparkasse Dieburg.

(Q): What was the reason and the goal of this project?

(A): The primary goal was the pending merger of two offices in Dieburg. Furthermore, to analyse and optimise the "entrenched processes", which were each put to the test by internal and interdisciplinary teams. This spatial change led to the creation of a new and neutral office space, i.e. to form two separate locations into one. In addition, the goal was to increase customer satisfaction, as well as employee satisfaction, while minimizing turnover and sickness rates.

(Q): Was there a model for the construction of this feel-good branch?

(A): The pioneer in terms of "feel-good savings bank branches" is Sparkasse Hamburg. Through the existing network, we were able to query the principle of implementing "feel-good factors".

(Q): How were the teams composed and did you receive support from top management, including financial support, due to the more intensive depth and length of planning?

(A): The duration of the pre-planning phase was 9 months, which is an indication of proper and all-inclusive pre-planning. Participation was voluntary. Teams were formed with the staff, across hierarchies, age groups and genders, and a project manager was also appointed. In Hamburg, customers were also included in the planning teams.

Structured solution-finding processes were also used in the preliminary planning, which gave us the chance to think in a user-oriented and user-related way. Furthermore, change management was carried out with the interior designer Verena Pankoke. The people involved in the project were very committed, which can be attributed to the good and open attitude of the employees and the quality of the change management carried out.

The entire project cost 750,000€ for an area of 450m^2 with approximately 14 employees (plus 1–2 trainees). It started with preliminary planning in mid-2016 and was completed in spring 2019.

(Q): What are the most important attributes in relation to the customer?
(A): It was a piece of hard work to align and agree on the essential attributes. In the end, we settled on the following attributes:

- Trust
- Community
- Helpfulness

Another challenge was identifying the well-being factors (Fig. 3.14).

The aim was to involve and find a good neighbourhood relationship, for example the "neighbourhood table" with free coffee, tea, water, newspapers and magazines. In addition, a changing art exhibition is organized.

The neighbours or customers can also sit down in retreat corners and read or just "chill out" (Fig. 3.15).

According to the concept, two office segments were defined, on the one hand the public area where employees and customers can interact, and on the other hand the area planned exclusively for employees (Fig. 3.16).

Conclusion
The space is very well adapted to the users (employees and customers). The well-being of the employees has increased, as shown by an internal survey.

Well-being Factors from the Service Provider Sector: Savings Bank Branch
Newly added factors are marked in bold.

- Architecture: functional room layout
- Health: place of rest
- Work-life balance: change management, appreciation
- Communication: **Constructive measures,** informal communication, communication concept (community)

3.6.2 Fraport

Fraport AG is one of the leading international companies in the airport business and operates at 30 airports on four continents. In 2018, the Group generated an annual result of around 506€ million on revenue of 3.48€ billion. In 2018, a total of more than 176 million passengers used the airports with a Fraport share of more than 50 percent. [. . .] At its home base in Frankfurt,

Fig. 3.14 Communication node: the neighbouring table

> Fraport welcomed more than 70.5 million passengers in 2019 and handled a cargo volume of around 2.1 million tonnes. [. . .] Frankfurt Airport City is also the largest local workplace in Germany, employing close to 81,000 people in around 450 companies [. . .]. This makes Frankfurt Airport in the middle of Europe one of the most important air traffic hubs in the world and an important infrastructure for Germany as a business location.[9]

The interview was conducted with Mr. Patrick Schäfer, Head of Real Estate Management at Fraport. The interview relates to administrative buildings and lounges.

(Q): Fraport's core business is a large area of responsibility, how would you describe it?

(A): We are currently working on two topics in particular—Terminal 3, which is currently under construction, and the further development of the existing terminals, especially with regard to the airport lounges. As a globally active airport operator, we can draw on many years of experience in this area. The focus is always on the passenger experience and well-being.

[9] Fraport (n.d.).

Fig. 3.15 Retreat option

Terminal 3, one of the largest infrastructure projects in Europe, is currently being built in the south of Frankfurt Airport. The new building will score with an attractive feel-good concept: From a large-scale glass façade to appealing quiet zones and special light installations to guide visitors and set the mood (Fig. 3.17).

But we are also constantly working on modernizing and improving our infrastructure in the existing terminals. For example, we have been offering our passengers a terminal hotel in Terminal 1 since 2017. Passengers can stay overnight in the transit area of the airport without leaving the security area. The rooms are specifically designed to meet the needs of passengers with a short length of stay. Guests can book the rooms flexibly, whether for a few hours for "power napping" or for a full night.

As a further convenience, passengers can also spend their waiting time before departure in one of the airport lounges. There are several lounges that are available to all passengers regardless of airline and booking class (Figs. 3.18 and 3.19).

Author's note: The lounges can be described as rather "mundane" or not very innovative with regard to well-being factors. According to the current state of knowledge, the following well-being factors should be reconsidered for lounges:

Fig. 3.16 Workspace for employees according to the shared desk principle

- Sport while relaxing (table football game, table tennis)
- Power napping
- Colours
- Plants
- Quiet zones, communicative zones
- Background music
- Water sounds
- Healthy building (ceiling height).

(Q): What is a well-being factor for you?

(A): Plants and furniture are clearly among the classic well-being factors. The acoustics and the colour concept are also determining elements in the room design.

Fig. 3.17 Relaxation with a view

Fig. 3.18 The LuxxLounge at Terminal 1 of Frankfurt Airport

Fig. 3.19 The Sky Lounge at Terminal 1 of Frankfurt Airport

(Q): How did you determine the feel-good factors?

(A): We have developed many factors based on our own experience. At the same time, we are part of a good network with other service providers and can use this to obtain new impulses. We are also in regular contact with companies that are moving into new spaces in order to benefit from their experience.

(Q): Who pushes the consideration of well-being factors?

(A): All areas are represented in the preliminary planning, from architects to colleagues in room and furniture planning. This ensures that all aspects are taken into account. The aforementioned team then jointly decides on the relevant well-being factors.

(Q): Were structured solution-finding processes also used, such as mind maps, brainstorming, programming, etc.?

(A): Tools such as mind maps are now part of our daily work. For example, we are increasingly using agile methods such as Design Thinking or Scrum to solve problems. The advantage of interdisciplinary teams is obvious. We benefit from a variety of different perspectives and can learn from each other.

(Q): Are the costs higher than for a conventional project?

(A): Although the cost of equipping the room with innovative technology may be higher, this investment pays off again thanks to conformity with user requirements and better performance.

(Q): What "profit" are you counting on?

(A): Our goal is, of course, satisfied customers, but well-being factors also mean that the specific key figures for employees, e.g. the fluctuation rate and lower sick leave, are good. In addition, a performance boost is expected.

Conclusion

Setting up or revitalizing lounges is usually an important building block for customer satisfaction and can be described as a significant success factor in most cases. In order to achieve customer acceptance of the lounges, the concept and design must be renewed more frequently. The design should move away from conformity, this increases the level of awareness because it is talked and written about.

Since the service profile is very broad, a wide range of well-being factors can be named.

Feel-Good Factors from the Service Provider Sector: Airport

- Architecture: functional room layout
- Perception: acoustic concept, light and colour concept
- Nature: plants
- Health: rest areas, power napping

3.7 Urban Development

Factors that influence well-being also exist in the surroundings of buildings. Here, the focus is primarily on issues relating to architecture and the choice of colours in façade design, but there are also other factors that determine why we feel comfortable in a city.

The interview was conducted with Professor Dr.-Ing. Christine Kohlert, Managing Director of the Drees & Sommer project management office.

(Q): What is the meaning of the term "well-being"?

(A): Psychologists, economists, philosophers, and other social scientists use the term well-being as an umbrella term to describe the mental, physical, and social state of an individual or group. Sometimes well-being is high, meaning that the state of the individual or group is positive; when well-being is low, things are generally not going well, and the state is called negative.

It is important to make a distinction between the strictly physical notion of well-being that a doctor might use and the broader notion that we use here in the context of space, or more specifically, urban space.

If you look at the different factors when a person feels good, it has a lot to do with appreciation, but also with motivation and creativity and, last but not least, with a positive environment in which he moves. Of course, this environment does not only mean the effect of a building inside, but also the charisma of the built environment seen from the outside, as well as the entire urban environment and beyond that green spaces and nature in general.

(Q): What factors influence well-being?

(A): The environment plays an important role in human well-being. In 2012, psychologists James K. McNulty and Frank D. Fincham expressed the idea that well-being "is not determined solely by people's psychological traits, but jointly by the interplay of those traits and qualities of people's social environments."[10]

Nature and open spaces play a particularly important role here. In 1989, the environmental psychologists Rachel and Stephen Kaplan investigated how creative thinking can be stimulated in their Attention Restoration Theory (ART). They found that after spending time in nature, people were able to concentrate better. They postulated that what they called the "soft factors" of the natural environment—rustling leaves and flowing streams and drifting clouds—created a kind of "effortless attention" that greatly improved the flow of ideas. While "communing" with nature, you feel freed from stress and negative thoughts, and feel like you can think freely, be creative, and feel good here.[11]

(A): Are there also influences in the planned architectural elements?

(A): Inner courtyards, open spaces between buildings or atriums can also have a similar effect. The British Museum in London's Bloomsbury district is not only distinguished by huge exhibitions with around eight million works from all continents, it is also famous for its architecture, not least for its Great Court with its impressive glass roof by Norman Foster. Conceived as a new entrance to the museum, it very quickly became a piazza, a square, a town square—the indoor equivalent of a large, open public space used for community gatherings and as a meeting place to chat or meet before going into the museum, or where you can just enjoy this beautiful space. It is a busy place with shops and a café, but for visitors the hustle and bustle is not disruptive, they come to relax but also to work there. The conscious design as an entrance has given way to the unconscious use as a space that supports aspects of what people need to think creatively and feel comfortable. This pleasant feeling—the feel-good factor—is also supported by the natural light flooding into the space

[10]McNulty and Fincham (2012), pp. 101–110.

[11]Cf. Kaplan and Kaplan (1989).

and the pleasant natural materials. This combination of materials, curves and natural light represent elements of biophilic design that bring nature into the space, including the urban space, and can be experienced by all users of the space, providing a natural environment.

(Q): Can you give another example of this?

(A): In Riehen, a suburb of Basel, the Fondation Beyeler Museum does something similar. The building is set in an English park and looks out over the vineyards and grain fields of the Tüllinger Mountains at the foothills of the Black Forest. The marriage of nature, art and architecture was a specific aim of the museum's patron, the art dealer Ernst Beyeler, who commissioned the multi-award-winning Italian architect Renzo Piano to design a new space to house his own collection of paintings and sculptures—including a famous work by Monet that fits seamlessly into the generous surrounding space. The intention was always to flood the museum with natural light, and the bright glass roof allows this to happen. Natural light enters every room of the museum through the glass roof, which almost appears to float. Frosted glass keeps shadows away from the exhibition spaces. A winter garden with seating offers views of the green beyond the museum walls, which link the architecture and nature of the building by extending into the surrounding park. All these elements help to complete the human interaction with nature that is so precious both inside each room and outside in the spatial environment of the building.

(Q): How do I get a good vibe?

(A): Good mood matters. There is a lot of research that shows that a good or positive mood promotes well-being and helps us come up with more good ideas. When we are in a better mood, we are more likely to get along with others, think rationally, be healthier and also think creatively. This is not just an individual phenomenon, but one that applies to groups of people as well.

Space and urban space have a significant influence on our well-being as humans, because space influences us emotionally. It therefore stands to reason that we consciously design spaces, including urban spaces, in such a way that they influence us positively. Feeling good promotes creative thinking and working, whether alone or with others: This means, for example, more innovation, improved problem solving and decision making, more flexible, thorough and efficient thinking on topics that are meaningful or interesting to the thinker, strategic thinking, constructive and cooperative negotiation, increased helping and interpersonal understanding, constructive suggestions and improved self-knowledge, and much more.

(Q): Do other factors exist in the context of well-being?

(A): Danish urban planner Jan Gehl is one of the most important advocates for cities for people. For him, it is important that streets, squares, entire neighborhoods are designed for the benefit of the users. He looks at cities on a small scale in detail to change inhospitable urban landscapes to fit the human scale and truly become cities for people. Urban space, instead of being experienced from a moving car or from above, must be experienced at the speed of a pedestrian to make them cities for people. This is the only way to achieve cities of coexistence for a better quality of life that puts people at the centre. It is always about the interplay between the physical environment and the activities in the public space. This also includes social activities in public space, which are an essential part of this interaction and for which the right spaces, squares and places must also be available. Only in this way can a social life of the most diverse activities develop. In this context, the quality of the individual segments of the public space plays an important and essential role. People should enjoy life in public space, chance encounters should be supported and people should once again be in the foreground. People want to experience other people, life, dynamics, joie de vivre and diversity. The design must not only support this, but really generate it.

So what contributes to this sense of well-being in the urban environment? A key factor in this is encouraging people to stay and spend time in the city. Jane Jacobs, journalist and neighborhood activist, observed as early as the 1960s how important the streetscape is as a feel-good factor in a city, she said: "If the streets of a city are interesting, then the city is interesting."

In order to ensure the highest possible quality of stay for a broad and mixed population, it is important to enhance public space. Jan Gehl has been working since the 1960s on quality criteria such as protection, comfort and enjoyment as feel-good factors.

Protection means high traffic safety and clarity for pedestrians. A variety of uses in the area, such as homes, offices, shops and restaurants, as well as good visibility, make people feel more protected and satisfied. Overlapping uses, good lighting, and a vibrant environment ensure better safety. Protection is also an important factor for good public spaces because of undesirable influences such as noise, stench or negative weather effects such as rain, snow, cold and heat.

Comfort means that good public spaces offer space for different activities and uses. It is important that both active and passive existence is made possible. Pedestrians should be able to move freely, but also be given the opportunity to stop or sit down in pleasant places with good surfaces, for example. These attractive places allow to observe, to have good views and to pursue different activities, also according to the season.

For people to really enjoy their stay in public spaces, the local assets must be exploited. A good public space is characterised by the fact that it is on a human scale and designed for people. These places allow people to enjoy the positive aspects of the climate, they offer a beautiful view and offer sensory impressions. The dimensions of buildings and public spaces are designed to suit people and allow, for example, alternation between sun and shade and the proper use of cold and heat, which is enjoyable for people. This also includes protection from wind and rain, but also good sights and the right vegetation.

Fig. 3.20 Valparaiso, Chile, graffiti

(Q): What is the relationship between colour in facade design and feeling good?

(A): There are very different examples of house painting and also very different reasons for it. Very successful graffiti can often be found in structurally weak regions, on the one hand to give the citizens there a voice and on the other hand to create identity in the neighborhoods. A good example is Valparaiso in Chile, the inhabitants there are supported, also get paint and spray cans to design and paint their houses and apartments, partly with graffiti, but also to paint their houses in different colors. The result of these actions was a much cleaner neighbourhood, less crime and happier residents (Figs. 3.20 and 3.21).

Even in our latitudes, well-designed facades signify appreciation for a neighbourhood, both for residents and visitors. Art actions on facades also show appreciation for people, such as Goethe and Schiller poems on Weimar's facades.

Successful facade designs can also be found in working environments, such as the Höchst Industrial Park in Frankfurt, where the colour designer Friedrich Ernst von Garnier painted industrial plants and chimneys in pastel shades to make the working environment more pleasant and appreciative for employees. Garnier let himself be guided in all his designs by his own theory of colour design:

> The creative forces of a society should make the working landscapes beautiful much more than the leisure landscapes. A person spends much of his life in his workplace, and it is my goal to

Fig. 3.21 Valparaiso, Chile, house paintings

make the working landscape natural, simplified, and soothing. This is possible through color.[12]

Plastics manufacturer Covestro in Brunsbüttel has also taken a similar approach and had a chimney designed by graffiti artist Martin Heuwold to express community and cohesion, as Covestro manager Patrick Thomas and site manager Steffen Kühling explain:

> The smokestack is no longer just seen as an industrial building, but also as a work of art. A great signal for Covestro and the Brunsbüttel site. [. . .] The employees here and also the people in the surrounding area will identify with the new smokestack and be proud of it because it is so colorful and, above all, unique. [. . .] Many things have already fallen into place after Covestro became independent on September 1, 2015, some still have to find their way. We use the symbol of the puzzle piece when cooperation, cohesion and unity are to be expressed. That's how the idea for the smokestack pattern came about.[13]

But colour is not the only way to achieve a sense of well-being in urban spaces; greenery between and on top of buildings, green spaces and places to stay are also essential factors for a vibrant city where people enjoy spending time, are inspired and motivated, and enjoy working (Fig. 3.22).[14]

[12] Infraserv Höchst (n.d.).

[13] Chemical Engineering (2016).

[14] See Sect. 3.8.

Fig. 3.22 Milan, Bosco Verticale by the architectural firm Boeri Studio

In Milan, in the Porta Nuova district, two green residential towers were built by the architectural firm Boeri Studio. Previously, a working-class residential area was located here. The center of the new development was left as a green space and recreational area. The project was about using urban space as well and effectively as possible and not overdeveloping it, thereby also improving biodiversity in Milan. The planting of the towers created new habitats and feeding areas for insects and birds; thus, the residential towers can act as biotopes between the public parks, the avenues and inner-city wastelands. The trees and plants on the façade will improve the microclimate in the apartments and on the balconies; the plants will mitigate noise, dust and heat. This will improve the quality of life for residents and provide a link to nature in an urban environment. “The Bosco Verticale is a unique architectural experiment in the world, a model for the inner city of the future,” says its architect, 57-year-old Stefano Boeri.[15] These high-rise buildings change the image of the city and increase the feel-good factor for people in a positive way. The greenery in front of the apartments not only ensures better air quality, but also serves as protection against cold and heat.

Ultimately, very similar factors apply to an urban space in which people feel comfortable as they do inside buildings. Based on research findings, 6 dimensions of well-being can be defined for humans: optimism—mindfulness—authenticity—belonging—meaning—vitality.[16]

Optimism is about inspiring joy in the present and discovering possibilities for the future. It is about some choice and control over how and where you move around the city, about choice, but also about appropriate transparency so that you can see, but also be seen, which has a positive impact on safety in the city.

Mindfulness allows one to be fully present in the moment and aware of one's surroundings. It is about creating urban spaces that help people connect with others. These are different places that allow you to control stimulation and are built with good materials that also have a calming effect on people, for example plants and soothing colours and healthy materials.

Authenticity means the freedom to be whoever you want, but this presupposes spaces in which you feel part of the whole, i.e. urban spaces in which you enjoy spending time and with which you can identify. This also means informal areas with a high quality of stay.

Belonging allows you to be connected to other people. This is also about designing a city in such a way that it not only supports people to reach their destinations well, but also to find places with good quality of stay where you can meet and exchange with others. Here again, it is important to have freedom of choice and different options where and how to move.

Meaning is promoted by a city that is designed to support people in their actions. The sense of meaning is promoted by making it possible to experience the culture, history and values of a city and the society in which one moves. A city needs places where this history can be read, and thus also a good and healthy mix of old and new.

Vitality needs the appropriate support from the different urban spaces. The sensory stimulation as well as different possibilities to perceive nature, different light and climate conditions, for example through walkways, and vistas stimulate people to be active and vital themselves.

Conclusion

Good places help people to engage themselves positively and thus they also support their own well-being and contribute to a healthy coexistence and a good community. Good urban design creates well-being for citizens and visitors.

---

[15] Nagel (2014).

[16] See Kohlert and Cooper (2018), p. 61 and Sect. 1.3.

Well-being Factors from the Field of Urban Planning

Newly added factors are marked in bold.

- Architecture: **healthy building and maintenance,** functional room layout
- Work-life balance: valuing users/citizens
- Perception: 4 L®, Colors
- Nature
- Communication: constructive measures, informal communication, communication concept (we-feeling)

3.8 Garden(s)

Interview with Dr. Michael Henze, Bundesverband Garten-, Landschafts-, Sportplatzbau e. V. (Federal Association for Gardening, Landscaping and Sports Grounds).

(F): Are gardens still modern?

(A): The Germans' favourite activity is gardening with 30%, followed by shopping in second place. The garden is a scarce commodity these days, because it demands what has become most precious in our society: time, attention and space. But a quality garden is worthwhile, because a well-maintained garden can add significantly to the price of a property.

Leading garden designer John Brookes in the 1960s said that a garden is like an outdoor room. Gardens are to be understood like extended rooms, which can be decorated and designed or used for relaxation.[17] Modern lounge garden furniture today is indeed reminiscent of living room interiors with large sofas or loungers (Fig. 3.23).

The garden should not only offer places for sunbathing, but also have places for shade. Without shade, cooling is not achieved, as evidenced by the difference in temperature between the city and the suburbs or forest of up to 10 °C. Dense buildings reduce the speed of the wind, so that heat is generated in the city, stored in the houses and slowly released at night.[18]Gardens or parks in urban areas can support this temperature balance.

In addition to the temperature, the air velocity and humidity also serve as a measure. These three parameters must be considered together so that the climate can be well regulated. The leaves of the plants provide fresh air, because they produce oxygen and bind CO_2.

[17] Cf. Young (2012), p. 190.

[18] Cf. Frey (2019), p. 58 f.

Fig. 3.23 The garden as an outdoor room

(Q): What kind of gardens are there?

(A): There are now a lot of boundaries and combinations, but the two main forms in garden design are natural gardens and formal gardens (Fig. 3.24).

A semi-natural garden is home to butterflies and bumblebees, ground beetles and wild bees, birds and mammals, for which native wild plants provide abundant food. In a near-natural garden, the beds and paths are usually laid out in organic shapes and surrounded by lush greenery. The planting should be as varied as possible, flowering and yet not wildly jumbled, but laid out with a basic structure. A varied play of colours is created, for example, by roses, perennials and grasses, hydrangeas, lavender and other small shrubs (Fig. 3.25).

Austerity can also be beautiful. This is proven by formal gardens, which create a feeling of exclusivity, clarity and harmony with geometric shapes and linear structures. Formal gardens form an ideal combination with, among other things, the geometrically designed houses in the style of classical modernism. Important elements of formal gardens are artistically pruned woody plants. The best known of these is the very versatile boxwood.

(Q): How did you determine the well-being factors for gardens?

(A): Many people are stressed by everyday life. Well-being and feeling content is seemingly unattainable for many people. The flood of information at work makes it difficult to switch off. The garden is a welcome escape. When external pressures

Fig. 3.24 Near-natural garden

Fig. 3.25 Formal garden

Fig. 3.26 Blue and green

become too much, it makes sense to retreat and take some kind of time out. One's own garden or a garden shared with other people, e.g. in a hotel or a public park, can then become a haven of peace and an oasis of well-being. According to Japanese studies, being in the green and the blue results in a reduced stress level and normalizes blood pressure and pulse rate (Fig. 3.26).

So you can enjoy peace and relaxation and recharge your batteries. A stay in nature, in the garden, in the park or in the forest and between fields leads to relaxation.

(Q): At what stage of the project do well-being factors start to play a role in garden planning?

(A): From the beginning of the planning, with the creation of the user requirements, the factors are taken into account. For this purpose, the appropriate well-being factors are selected in the planning team, which should be interdisciplinary in composition, in addition to the client's representative, architects and the gardening and landscaping experts, depending on the subject matter, physicians, urban and landscape planners, sociologists, psychologists, physiologists and others.

Fig. 3.27 Planned meeting place also in the garden area

(Q): Are solution finding processes used in the early phase of a project, such as mind map, brainstorming or programming?

(A): In general, solution finding processes are applied, of course depending on the size, complexity, degree of novelty, etc. of a project.

(Q): Are the positive effects of a green living environment on human health also measurable?

(A): There are areas where we are already running statistics. There is a clear improvement in the statistics on sickness absence, staff turnover and well-being (Fig. 3.27).

Well-being Factors from the Area of Garden(s)

- Health: place of rest, motivation to move
- Proximity to nature
- Perception: fresh air
- Communication (increase of social contacts)

References

Bastian, A. (1999), Erfolgsfaktoren von Einkaufszentren: Ansätze zur kundengerichteten Profilierung, Deutscher Universitäts-Verlag, Wiesbaden.

Borkmann, V., Rief, S., Iber, B. (2013), FutureHotel Gastbefragung, Fraunhofer Verlag, Stuttgart.

Chemietechnik (2016), Covestro puzzelt sich einen farbigen Schornstein, https://www.chemietechnik.de/covestro-puzzelt-sich-einen-farbigen-schornstein/ [Zugriff am 26.05.2020].

Deutsche Bahn AG (2017), Presseinformation – Mit dem Ideenzug blickt DB Regio in die Zukunft, https://www.deutschebahn.com/resource/blob/331052/71cda3aa89ed1dd0b70f4a808d9e5b15/PI-Ideenzug-data.pdf [Zugriff am 10.06.2020].

Franke, M. (2007), Lokaler Einzelhandel und Shopping-Center – Eine Betrachtung zu den Auswirkungen eines integrierten Shopping-Centers, VDM Verlag Dr. Müller, Saarbrücken.

Fraport AG (n.d.), Über uns, https://www.fraport.de/de/unternehmen/fraport/ueber-uns.html [Zugriff am 26.05.2020], Frankfurt am Main.

Frey, A. (2019), Der Sommer, die Stadt und der Klimawandel, in: Frankfurter Allgemeine Zeitung, 25.08.2019, Seite 59.

Gruen, V., Smith, L. (1960), Shopping Towns USA: The Planning of Shopping Centers, Van Nostrand Reinhold, 1960.

Infraserv GmbH & Co. Höchst KG (n.d.), Industriekultur im Industriepark Höchst, https://www.industriepark-hoechst.com/de/stp/menue/der-industriepark-hoechst/besucher/industriekultur/#notification-close [Zugriff am 26.05.2020].

Kaplan, R., Kaplan, S. (1989), The Experience of Nature: A Psychological Perspective, Cambridge University Press, Cambridge, England.

Kohlert, C., Cooper, S. (2018), Space for Creative Thinking: Design Principles for Work and Learning, Callwey, München.

Lange, C. (2009), Vertikal strukturierte Einkaufszentren in Innenstädten, Fallstudien zur gebauten Realität in Deutschland, Dissertation an der Technischen Universität Berlin.

McNulty, J. K., Fincham, F. D. (2012), Beyond Positive Psychology? Towards a Contextual View of Psychological Processes and Well-Being, in: American Psychologie, 62(2), S. 101–110.

Meyer Werft (2020), Wir sind die Meyer Werft, https://www.meyerwerft.de/de/unternehmen/wir_sind_die_meyer_werft/index.jsp [Zugriff am 26.05.2020].

Nagel, B. (2014), Dieses Hochhaus ist ein kleiner Wald, https://www.welt.de/finanzen/immobilien/article134513421/Dieses-Hochhaus-ist-ein-kleiner-Wald.html [Zugriff am 26.05.2020].

Seiferlein, W., Kohlert, C. (2018), Die vernetzten gesundheitsrelevanten Faktoren für Bürogebäude, Springer Vieweg, Wiesbaden.

Tran, Q. (2019), Stimmt die Atmosphäre?, in: Frankfurter Allgemeine Zeitung, 08.09.2019, S. 55.

Young, C. (2012), Neues Gartendesign, Dorling Kindersley, München.

4 Creating the Target Matrix

Preliminary Remarks

In the individual sections of the last chapter, research was conducted into well-being factors in various industries and sectors. The results were obtained from interviews, literature or both. Furthermore, already implemented examples were included in the data source.

In many companies, only isolated measures are implemented, which are carried out more or less "coincidentally", i.e. in a rather unstructured manner. What is missing is the whole picture or a master plan. This will be presented in the following.

Figure 4.1 compiles all the success factors that could be identified from the research. The basic information developed from the investigation of the well-being factors in office and laboratory buildings, to which the other factors of the other sectors and areas were added. The assumption that it is irrelevant where the individual factors come from was confirmed. A number of the factors were found in several sectors or areas, but in some cases in a different form or with different emphases. If possible, the weighting of the factors should be determined in each case by means of a further, more in-depth and empirical research activity. In addition, some factors could have been assigned to several categories, but this was not done for reasons of clarity.

It is assumed that the target matrix obtained from the numerous studies of the various industries and sectors is initially available in its entirety. This does not exclude the possibility that further factors will be added in the future, or that some will be assigned to other categories or removed again. Based on the available data, the factors provide the basis for individual concepts of the well-being factors, even beyond the industries and areas studied.

The matrix consists of two parts: The vertical ray describes the seven categories, in the horizontal ray the factors corresponding to them are assigned.

The individual categories and factors are taken from the following sectors or areas:

W. Seiferlein, *Well-Being Factors for Different Industries*,
https://doi.org/10.1007/978-3-658-34997-4_4

Architecture	Healthy building and maintenance	Functional room layout	A parking space for each vehicle			
Ergonomics	Adequate furniture	Office and workshop furniture	IT equipment			
Work-life balance	Mutual agreement	Employers & Company Alignment/ Change Manage-ment	Apprecia-tion			
Perception	Air	Light	Noise / music	Body	Colors	Fragrances
Nature	Plants	Water/ sea	Animals			
Communi-cation	Constructive measures	Informal communi-cation	Communi-cation concept			
Health	Hygiene/ allergies	Safety/ quality	Place of rest/power napping	mental activity	Sport and movement	

Fig. 4.1 Target matrix of well-being factors

Architecture

- Healthy building and maintenance: Urban Planning
- Functional space layout: office and laboratory construction, shopping centre, cruise ships, service providers, urban planning
- A parking space is available for every vehicle: shopping centre

Ergonomics

- Adequate furniture: office and laboratory buildings, hotel industry, idea train
- Office and workshop furniture: Office and laboratory construction
- IT equipment: office and laboratory construction

Work-life Balance

- Mutual agreement: office and laboratory building
- Employers & Company Alignment/Change Management: office and laboratory construction, service providers
- Appreciation: office and laboratory construction, hotel industry, service providers, urban planning

Perception

- Air: officc and laboratory construction, hotel industry, shopping centre, urban planning, garden(s)
- Light: office and laboratory construction, hotel industry, shopping centre, service providers, urban planning
- Noise: office and laboratory construction, hotel industry, shopping centre, cruise ships, service providers, urban planning
- Body: office and laboratory construction, hotel industry, cruise ships
- Colours: Office and laboratory construction, hotel industry, service providers, urban planning
- Fragrances: Hotel industry

Nature

- Plants: Hotel-business, service providers, urban planning, garden(s)
- Water/Sea: hotebusiness, cruise ships, urban planning, garden (s)
- Animals: Office and laboratory construction

Communication

- Constructive measures: Service providers, urban planning
- Informal communication: idea train, service providers, urban planning
- Communication concept: hotel business, idea train, service providers, urban planning, garden(s)

Health

- Hygiene/Allergies: Office and laboratory construction, hotel business, cruise ships
- Safety/Quality: office and laboratory construction, cruise ships
- Place of rest/Power napping: hotel business, idea train, service providers, garden (s)
- Intellectual activity: Cruise ships
- Sport and movement: Office and laboratory construction, hotel business, idea train, cruise ships, garden(s).

For the determination of the individual well-being factors to be taken into account using the developed target matrix, the following process flow is proposed:

- Formation of one or more interdisciplinary user teams. A mixture of different personalities is appropriate, but attention must be paid to the number of "alpha dogs" so as not to distort the result. Result: User requirement
- Comparison of the target matrix with the user requirement. Result: Target matrix with suitable well-being factors
- Fixing the well-being factors in question in a concept that contains a master plan of the realisation phases. Result: Actual matrix of well-being factors
- Implementation and use of the selected well-being factors
- Review of the concept in a fixed rhythm for necessary changes.

5 Explanation of the Well-being Factors

Preliminary Remarks
In order to facilitate the application of the matrix, the target matrix is described and explained in detail below. This enables users to better understand the factors and make comprehensible and sustainable decisions. However, it should also be mentioned at this point that the application of a factor is largely a matter of interpretation, for example, in the hotel-business, instead of background music, another acoustic sound such as chirping birds, can also be applied.

5.1 Architecture

5.1.1 Healthy Construction and Maintenance

Health in real estate is measurable, plannable, feasible and payable. This has been successfully tested in many hundreds of projects with several thousand units. The most important criterion is the load of indoor air with pollutants, for which there are official criteria. For those who already build and maintain qualitatively well, it is only a short way to tested healthier building.

In addition to the question of air exchange in the room, the health quality of the building materials, equipment materials and furniture used with regard to the emission of pollutants into the indoor air is of great importance for the quality of stay.

For the use of sustainable materials, it makes sense to set up a contractual target agreement for defined pollutants between client and contractors. The selection of building materials according to health criteria should take place after consultation with the specialist

W. Seiferlein, *Well-Being Factors for Different Industries*,
https://doi.org/10.1007/978-3-658-34997-4_5

trade or qualified skilled workers, because they know the building materials and processing conditions.

The challenges for companies in the real estate industry are enormous. Government requirements, for example in terms of energy efficiency and sustainability in new construction and refurbishment, are strict. And almost every day it becomes more demanding to create cost-effective living space and generate a reasonable return at the same time. In this context, it is worthwhile to analyse the terms, because with the increasing attention of the building owner to health issues, marketing is increasingly using terms that are ambiguous.[1]

The introduction of new technologies may be high risk as they have not yet been proven through regular business use and may not deliver the required function or reliability, so they must always be evaluated and defined. Time and resources are also factors to consider that may count against a new building.

However, new buildings enable a quantum leap in terms of the technology (and knowledge) used because they challenge the creation of a future-oriented building design in which a long-term view can be taken into account. It can be more cost-effective to construct a new building that meets the needs of the building than to invest in stopgap solutions. With ongoing renovation and adaptation to new technologies, the limits of maximum possible improvement will eventually be reached. However, if the planning and design for a new building takes into account all available and suitable new technologies, it is possible to achieve a leap in innovation.

This approach was implemented by the architectural firm Grafton, whose new building of the Universidad de Ingeniería y Tecnología (UTEC) in Lima was awarded an exceptional example of civic architecture by the Royal Institute of British Architects in 2016 (Fig. 5.1).

In the distinct vertical structure, which accommodates the temperate climate conditions in Peru, green terraces with crevices, overhangs and grottos form the campus building and provide shade. Inside, a series of communication platforms and connecting bridges create meeting places with open and transparent circulation flows. Classrooms, laboratories and offices are enclosed, inserted into and suspended from the exposed concrete structure (Fig. 5.2).

This concept of a vertical campus and the mix of open and closed spaces is extremely unconventional, but makes the most of this building's communication, both visually and spatially.

5.1.2 Functional Room Layout

The design of spaces is a decisive factor for their users in terms of well-being. It is interesting that the rooms for production workplaces and offices were initially not designed

[1] See Bachmann (2018), p. 139 ff..

Fig. 5.1 Polytechnic in Lima, Peru: open, transparent and vertical

Fig. 5.2 Technical college in Lima, Peru

Fig. 5.3 Use of height for additional offices

by architects or designers, but predominantly by organisational consultants, engineers, mathematicians and computer scientists. The architect was only responsible for implementing the catalogue of requirements.[2]

It is particularly true for commercial workplaces that when offices are built, both the space can be used economically, but also process-related requirements can be met. This is the case, for example, at Škoda in the Czech Republic, where the height of the factory hall made it possible to build additional offices (Fig. 5.3).

Such a functional layout can improve communication by, for example, combining skilled personnel who were previously located far away from the production site. A mechanic or fitter is assigned to the line in a stationary state. If a malfunction occurs there, the specialists can be on site more quickly and rectify it.

The laboratory layout has also changed over time. Where there used to be a strict separation of the premises, today you find a mixture of office, lab and support that is relevant for the use.

The dimensions of the rooms influence the type and quality of the work done in them.[3] If, for example, the ceiling height is above average, the room is suitable for conceptual thoughts, lower rooms are ideal for concentrated work.[4]

In general, the layout of the space should be designed to support the six dimensions of well-being (optimism, mindfulness, authenticity, sense of belonging, meaningfulness,

[2] See Rumpfhuber (2013), p. 49.

[3] See McCoy and Evans (2002), p. 409 ff.

[4] See Meyers-Levy and Zhu (2007).

vitality).[5] This requires a mix of open and enclosed spaces that take into account the physical, cognitive and emotional needs of those working within them. This can be achieved through a system of interlinked zones for focused individual work, teamwork, recreation and social interaction. Spaces should be able to be personalized and customized and avoid rigid workplace standards by allowing for standing, sitting, and walking work. A transparent space layout allows a relationship of trust to develop because people can see each other and be seen. Rooms that have a calming effect through materials, surfaces, colours, light and views, and areas that allow employees to decide for themselves how many and which stimuli they want to perceive through their senses, and to withdraw if necessary, increase well-being.[6]

For a barrier-free design of the premises—not only of workplaces for severely disabled people—attention must be paid to the condition of the floors. They must not contain any tripping hazards, must be slip-resistant, level and easy to clean.

There should also be enough storage space available in the office rooms. This can be realised by group storage near the associated workstations. A central changing room or cloakroom with lockers to supplement the coat hooks at the workplace and the visitors' cloakroom is also useful.

For reasons of noise protection, copying machines and printers are to be accommodated in separate copying rooms, if possible. In addition, the kitchenette is to be planned as a multifunctional room in which meetings can also be held and visitors received in order to increase the communicative aspects of this room.

5.1.3 A Parking Space for Each Vehicle

Especially in city centers, it is not yet common practice to provide employees and customers with a parking space, and the focus is still increasingly on individual motorized transport. In the Smart Economy, however, a new way of thinking is increasingly taking hold that focuses not on the car as a product but on mobility as its effect, i.e. mobilization without individualization.[7] This is reflected in the trend towards car sharing and the expansion of public transport. In this context, it is also important to network existing tools and support the development of new transport systems. Alternative bicycle parking facilities with roofing and possibly a charging station for e-bikes in sufficient number and quality should therefore be considered. This measure would be perceived by customers as a feel-good factor.

[5] See Sect. 1.3.

[6] See Steelcase (2013), p. 21 ff.

[7] See Oetter (2015).

If parking facilities for bicycles are provided in the operational context, the desire for changing facilities, showers and lockers etc. will also soon arise. For the employees, this would be an expression of appreciation, which is currently offered rather rarely.

5.2 Ergonomics

5.2.1 Adequate Furniture

Ergonomic furniture is easy to adapt to different body sizes, needs and preferences and allows movement in the workplace. High quality furniture works flawlessly, is easy to use and lasts a long time. A modern, light wood decor and white finish is not distracting and lifts the mood.

Mattresses and upholstery for beds, loungers, lounge and rest furniture must not have any harmful vapours, should be ergonomically designed to be easy on the back and adaptable to different postures. For individualisation, fabrics and covers, which should generally have as pleasant a feel as possible, can be provided in a choice of colours and patterns. In addition, there are also fabrics that increase blood circulation and dilate the blood vessels, allowing more oxygen to reach the cells and making you feel more energetic and alert.[8]

Cabinets and storage areas can also be used as room dividers, thus offering the possibility of separating different areas from each other in a soundproof manner and at the same time turning them into places of encounter. They should offer sufficient space to systematically store working materials and documents. Lockable compartments ensure privacy and the safe storage of personal items.

5.2.2 Office and Workshop Furniture

Desk work challenges the body in many ways. Sitting for hours without moving stiffens the postural apparatus, concentrated staring at the screen reduces eyelid blink frequency and tear fluid, which dries out the eyes and causes them to burn, and equipment that cannot be individually adapted to the body leads to poor posture, muscle tension, headaches and ultimately to avoidable illnesses and absenteeism.

An ergonomic swivel chair is height-adjustable, has castors and adjustable armrests, a horizontally and, if possible, vertically adjustable seat, the backrest can be moved vertically and the hardness of the backrest suspension can be individually adjusted. If possible, the desk is also height-adjustable and equipped with a cable duct to increase safety. The chair

[8] See Steelcase (2019), p. 98 ff.

and desk should be adjusted so that the arms on the desk surface form a right angle to the upper body when operating the keyboard.

In order to increase the possibilities of movement, alternating working positions should be offered where possible, i.e. standing, sitting and lying down.

For the furnishing of different areas, the following variants with the corresponding equipment are possible[9]:

- Classic workstation areas: sit-stand desk, electrically height-adjustable with display for height indication to simplify adjustment; screens to inhibit sound propagation as well as for sound absorption and as protection against visual disturbances; ergonomic office swivel chair
- Short-term workstations (touchdown workstations): electrified solution for up to a maximum of eight people with acoustically effective screens and, if necessary, wireless charging; ergonomic office swivel chairs with intuitive adjustment options
- Workshop rooms: flexible tables, foldable and/or on castors as well as plug and play electrification; flexible chairs (light weight, on castors); smartboards or whiteboards; storage facilities for workshop materials
- Creative or project spaces: table at standing height; standing benches or standing aids; smartboard/monitor with wireless connectivity; whiteboard.
- Rest areas: comfortable, attractive and clean relaxing and lounge furniture
- Cafeteria/bistro: varied, attractive seating at catering level; different table variants (different heights) and respective number of seats

The furniture for the workforce in the workshop or production should be determined by the functions of the actors and the hierarchy. The production area consists of the following groups of people: Plant Manager, Plant Engineer, Foreman, Operator, Mechanic, Electrician, Line Worker, etc. The hierarchy levels in a production plant up to the foreman require an office, whereby in principle any form from a cubicle office to a combination office with shared desk is possible here.

The furniture used in production often depends on the type of production, so that need-based seating options must be created that can withstand the different requirements in terms of mobility, space requirements and ergonomics. Practical steel furniture is the preferred choice for production, crafts and construction sites.

5.2.3 IT Equipment

The range of ergonomic keyboards and mice has become increasingly extensive in recent years, making it possible to meet a wide variety of needs in terms of hand size and operator options. For workstations occupied by only one person, this individualisation is not a

[9] See Steelcase (2013), p. 38 ff. and Steelcase (2019), p. 58 ff.

problem, but for shared desks, compromises have to be found. Here, either the lowest common denominator could be decisive, or different versions are procured if several of these workstations are to be set up.

The monitor should be adjustable horizontally and vertically if possible and should be positioned so that it can be viewed directly without having to turn the upper body. The monitor should be calibrated by an expert and adjusted in terms of brightness and contrast to the individual needs of the user. To relieve the eyes, the prescribed visual aids must be used and, where necessary, regular relaxation exercises must be carried out.

In addition, equipment should be available for better communication, which must also be intuitive to use by people who do not use it so often.

5.3 Work–life Balance

5.3.1 Mutual Agreement

A mutual agreement lays down rules of conduct for personal dealings with each other and specifies, for example, how to communicate with each other; what measures are to be observed with regard to order and cleanliness at the workplace, or whether and under what conditions a pet is allowed in the workplace. It is required where massive changes are planned. It is important that these agreements take people's well-being into account and that the agreed code of conduct is adhered to by all employees from trainee to management level.

For defining a mutual agreement, the six dimensions of well-being from office building can provide guidance: optimism—mindfulness—authenticity—belonging—meaning—vitality.[10] Ultimately, it is about also making the company in which one works a place worth living in, which respects each individual with their needs and potentials and supports them in developing their possibilities.

5.3.2 Employers & Company Alignment/Change Management

In the context of an Employers & Company Alignment, a company wants to adopt a new orientation (alignment) of innovation performance and process optimization. Its implementation depends on the actors, i.e. the people. Therefore, employees and management must be aligned to the goal before implementation. This includes a positive external image, which is of crucial importance in addition to professional competence. However, if you want to stand out from the crowd in order to create your own corporate identity, you should leave the antiquated ways in order to manifest the new orientation for the employees and

[10] See Steelcase (2013), p. 21 ff. and Sect. 1.3.

the management, i.e. also the company in a sustainable way. To do this, it is necessary to develop a coherent culture and vision that takes into account the specifics of the company and its workforce.

The development of a vision requires a professional change management program with which the company management involves its employees in the process. The possibility of participation promotes the acceptance of the employees for the changes. Comprehensive project communication also gives them a sense of security and integrates current information needs. In this way, the desired corporate culture can be anchored throughout the entire change process and in the design of the new environment.

5.3.3 Appreciation

Job satisfaction is a significant factor influencing the physical and mental health of employees and thus the success of the company, because it is expressed in lower sickness rates and higher productivity. A management and performance culture based on appreciation, respect and qualified communication contributes to job satisfaction.[11]

Appreciation, however, consists of more than praise and recognition, but encompasses "an attitude of heart and mind that always sees the person and not just their output."[12] This means that the value of an employee can not only be measured by his performance, but also his whole personality with his abilities and weaknesses.[13]

Genuine appreciation is characterised by the fact that it is addressed directly to the person concerned, who can thus also feel that he or she is really meant, and that it specifically names which achievement or quality is being recognised. Appreciation can be conveyed through open and respectful communication at eye level, seeking the opinion and advice of the other person, through constructive criticism, recognition of needs, giving of time and attention, small gifts and leadership with trust.[14]

5.4 Perception

Perception is everything that is felt or noticed with the five senses eye, ear, touch, nose and taste, i.e. stimuli that have an influence on the person from the outside world. The visual channel is most involved in the processing of information, accounting for 70–80%.[15]

[11] See Martin and Schlipat (2014), p. 4.

[12] Mai (2016).

[13] See Mai (2016).

[14] See Mai (2016).

[15] See Seiferlein and Woyczyk (2017), p. 80 ff.

In the private living area as well as in the commercial working area, the well-being factors can be defined by the expression 4Ls®, which include the areas of air, light, noise and body.[16]

The well-being factors belonging to perception are the most critical compared to the other factors found, because people's perceptions are very different.

5.4.1 Air

When there's a draft, when it's too warm, when it's too humid, facility management is often called in immediately to remedy the situation. If one wants to grasp the reason for the justified complaint, one inevitably comes across the numerous alternatives for room air generation. No matter in which direction the sensation tends (too hot/too cold), it is important to aim for only moderate deflections when correcting the climate setting. The extreme absence of a parameter also affects health, e.g. the influence of dry air has negative effects.[17]

In order to avoid these negative effects, effective control of ventilation equipment is necessary, e.g. opening windows for night cooling or in the early morning hours.

Dry room air should be avoided, the lower limit of the comfort zone is between 40 and 45% humidity, but above 60% humidity the risk of mould growth increases. Weekly cleaning of the ventilation systems counteracts the risk of germs.

5.4.2 Light

To avoid the negative effects of incorrect lighting, the light should be part of a color concept. Too little light lowers the level of mobility of people and their feeling of being happy, and leads to illness and isolation, furthermore depression and stress can occur.[18]

Optimal for maintaining health is a workplace with daylight and a view. Dynamic lighting conditions, as they occur in the course of a day, can reduce fatigue in people, and employees retain a fresh feeling.[19] Workplaces should therefore be illuminated by lamps with light colour adapted to the time of day (Human Dynamic Light), which is already available in office buildings. In production, it should be examined whether their use represents a feasible alternative.

In general, the entire room should be evenly lit, whereby the light must be able to be switched individually. Glare or reflections due to daylight incidence should be provided

[16] See Seiferlein and Kohlert (2018), p. 31 ff.

[17] See Hahn (2007).

[18] See Heidari et al. (2013).

[19] See Stefani (2017).

Fig. 5.4 Sound Shower for generating directional sound waves

with adjustable light protection devices, and blinds or curtains should be available and function properly against strong sunlight.

The lighting on the desk should be adjustable in steps and located to the side of the monitor or directly above it on the ceiling.

5.4.3 Noise

Noise is one of the most significant factors of the 4Ls®. Therefore, the required absorption rate for the office or a commercial workplace with good noise insulation to the inside and outside should already be determined in the planning stage. It should be taken into account that devices such as printers, copiers, scanners, faxes, blinds and air conditioning are noise-reducing.

In an open-plan office, where conversations and discussions are potentially taking place, the level of background noise generally rises to at least 45 dB, resulting in a drop in employee performance. However, the background noise level should never exceed 48 dB.[20] For professional demarcation, the so-called Sound Showers have therefore often been installed, the aim of which is to generate directional sound waves that are no longer audible in the neighbouring area (Fig. 5.4).

With the Sound Shower, it is possible to set up multifunctional zones that are acoustically and visually shielded from the respective adjacent workplaces and serve as a place for concentrated work, bilateral meetings and telephone calls.

The relationship between the loudness of a soundscape and creativity was investigated by Mehta et al. who found that people exposed to moderate volume conditions may have more creative ideas than those exposed to very low or very high volume.[21]

[20] See Roelofsen (2008), p. 202 ff.

[21] See Mehta et al. (2012).

The study by Haapakangas et al. also showed that a distracting conversational backdrop can be masked by defined background noises, but that different reactions are shown depending on the type of noise. For example, vocal music has the same effect as the spoken word in decreasing productivity. Instrumental music has medium values, while sounds from nature, such as spring water, are best at masking disruptive conversations in open-plan offices.[22] In a wide variety of industries and sectors, background music is already regarded as a success factor, e.g. for the willingness to buy products, or is used for relaxation. In open-plan offices, the realization of background music would be possible from the technical side via the electro-acoustic system available in every company, but Haapakangas et al. recommend constant sounds for the masking of conversations.[23] Suitable locations for the use of background music would therefore only be jointly used rooms such as stairwells, foyers and toilets.

In a production plant, on the other hand, the existing noise level caused by machines and plant components is often at the permissible limit anyway, so that neither background music nor noise masking can be used here.

5.4.4 Body

The body factor refers to all things that can be eaten and drunk. This raises the general question of how the food supply for employees is organised, i.e. whether it takes place externally or internally, i.e. whether there is a canteen in which a varied, possibly also vegetarian/vegan cuisine with healthy ingredients is offered, or whether there are cafés with a selection of healthy foods nearby.

Eating and drinking should preferably take place in a common room, a kitchenette or a multifunctional room. It contributes to well-being if, instead of chocolate bars, perhaps some healthy food such as apples or other local fruit is offered there. The in-house beverage stations should be spatially and acoustically separated and provide different beverages such as water, juice, coffee, tea.

5.4.5 Colours

Colours in light make the objective world visible to us. Matching the colours in a room to the lighting concept is therefore an important prerequisite for the design of pleasant

[22] See Haapakangas et al. (2011).

[23] See Haapakangas et al. (2011).

working environments. People see bright things first, which is why the places and surfaces that should attract the most attention should be designed brightly.

The example of the coloured painting of apparatus, pipelines or chimneys according to Garnier's concept shows that colours are a well-being factor. Studies showed that people have a higher stress level with red objects or in red rooms than in green or white room conditions.[24] A blue environment allows individuals to focus more on a task, but a red environment tends to divert attention away from the task.[25]

Warm colors are often associated with interest, communication and love, but also power and aggression. They have an active, exciting and invigorating effect on the viewer. Cold colors, on the other hand, convey distance, coolness, an impression of thinking, freshness and reason. The effect of cold colours is often also described as passive, calming, relaxing and refreshing.[26]

Greenery and natural materials enhance the mood and contribute to a sense of well-being and a good climate.

In addition, colours also have an influence on our perception of temperature. In a room painted blue-green, you feel cold at 15 °C, whereas in an orange room you only feel cold at 2 °C.[27]

It is very revealing how differently the effect of colours is defined. Trautwein therefore demands that a colour concept be created that manifests itself throughout and also takes into account the transition from the interior to the exterior areas.[28]

5.4.6 Fragrances

When one thinks of industry, production or workshops, one thinks of unpleasant and sometimes even harmful odours, where there is a need for action on the part of the employer. However, this is also the case with harmless odours, which also affect the well-being.

The University of Munich conducted a trial with Deutsche Bahn and found that customer satisfaction was higher on a regional train when a soothing scent was emitted.[29]

Odor evaluation methods are state of the art in industry. The spectrum of applications ranges from the detection of harmful vapours to the deliberate scenting of products to create a product-specific, acceptable odour sensation. The addition of more or less clearly

[24] See Kutchma (2003).

[25] See Stone and English (1998).

[26] See Nüchterlein and Richter (2008).

[27] See Braem (2009).

[28] See Trautwein (2018), p. 102 ff.

[29] See Hortig (2013).

perceptible scents into the air by stand-alone devices in the room is not necessarily purposeful.[30]

5.5 Nature

Several studies confirm a connection between green spaces and health. Even brief contact with nature or vegetation, such as wilderness, forests or bodies of water, but also with urban greenery, such as parks, generates positive feelings and moods, relaxation, well-being, happiness, etc., and reduces negative feelings, including aggression, fear and anger. In a Japanese longitudinal study of 3100 older Tokyo residents, survival after 5 years increased with the amount of access to green spaces or green walking opportunities, but also with community involvement. The correlation was particularly large for people with few physical deficits. Purely populated environments, on the other hand, lowered sentiment.[31]

5.5.1 Plants

Plants at the workplace and at home can contribute to the regeneration of forces. They reduce indoor climate-related complaints at the workplace (general malaise, fatigue, headaches, skin and mucous membrane irritations). Indeed, in Nieuwenhuis' study, enriching a formerly spartan space with plants and greenery led to a 15% increase in productivity—a figure that closely follows the results of previous laboratory studies.[32]

Air-purifying plants absorb few pollutants through their leaves, but much more through their roots. The conventional plant pot has minimal oxygen production. In contrast, the AIRY system alternative is equipped with a unique aeration system that guarantees optimal aeration of the root zone. The air reaches the roots via the chimney effect of the lamellas. The system can purify up to 40 cubic metres of air (equivalent to about a 16 square metre room) within 24 h. In the process, the plant converts the pollutants into nutrients that it can use without leaving any residue. This makes the plant eight times more efficient than in a conventional pot.[33]

A green working environment is consistently more pleasant for employees, increases concentration and thus causes higher productivity for companies.[34] In many areas, however, it is not possible to place plants, not only because they require water and care, but also

[30] See senkonzept GmbH (2017).

[31] See Brämer (2008), p. 5.

[32] See Nieuwenhuis et al. (2014); Henn and Meyhöfer (2003).

[33] See AIRY Greentech GmbH (2020).

[34] See Nieuwenhuis et al. (2014).

Fig. 5.5 Two plants communicate with each other

because contamination can occur between the plant and the pharmaceutical product in cleanroom areas, for example.

A workspace with a connection to nature is the optimal. A real outdoor green landscape would be welcome, but is not always feasible. As the examples of the Merckle and Seegmüller companies showed, pictures or photos can have the same effect as a real landscape. For this reason, it is desirable to attach a photo to each area that comes into question for this purpose.

In the example of Fig. 5.5, it can be seen how the two poles of inside and outside communicate with each other. This can create a smooth, imperceptible transition between inside and outside. The visual connection with nature can improve mental engagement and attention. Likewise, blood pressure and heart rate can be reduced. Furthermore, people or employees improve their positive attitude and general feeling of being happy.[35]

[35] See Aldwin (2007).

5.5.2 Water and Sea

When we enter the lobby of a hotel, we often hear—and a moment later see—a water feature. Without thinking, our brain anticipates a thought: the splashing of a small waterfall comes to mind, reminding us of last year's vacation. Thoughts continue to wander around the holiday and we prefer to think of the beautiful experiences. It is similar with thoughts of the sea, which create memories of experiences there: The wide horizon divides the sea from the sky. On the beach we can walk endlessly and the feeling of freedom arises. Both times we feel good!

The "water" factor includes all water features such as fountains, streams or the installation of a drinking fountain. This refers to all structures that produce the sound of splashing water, which, as explained above, contributes significantly to well-being and increased concentration. Water features are only suitable for production to a limited extent, but can be installed in rooms that serve informal communication.

Here, too, the view of water or the sea can be substituted by photos or images and thus create a good feeling. The technology of the Kasper company, for example, produces brilliant images that also have a space-expanding effect (Fig. 5.6).[36]

5.5.3 Animals

The positive effect of *animals* on the psyche of humans has long been known. Pets such as dogs also contribute to the physical health of their owners because they encourage them to exercise in the fresh air.[37] Various studies have also shown that having an animal in the working environment promotes motivation, that there is demonstrably less absenteeism in offices with dogs and that the sickness rate falls. Furthermore, it is positive to mention that the stress level decreases due to the eye and petting contact.[38]

Many companies that do not yet allow dogs believe that there is not much acceptance of this among employees. In contrast, the Office Dog Index 2019 found that 85% of respondents from companies without dogs are positive about this issue and even almost half of employees with their own pet would forgo a pay rise if they were allowed to bring their dog into the office. This could even increase loyalty to the employer, as the willingness to change is significantly lower among employees with an office dog. This would also be an opportunity for employers to increase their attractiveness to new applicants and retain them in the company.[39]

[36] See KASPER GmbH (2020).

[37] See Graen (2019).

[38] See Kals (2019) and Bundesverband Bürohund e. V. (2019).

[39] See Bundesverband Bürohund e. V. (2019).

Fig. 5.6 Image of beach and sea

However, before permission is given to bring an animal, it is essential that the conditions for doing so are set out in a mutual agreement.

5.6 Communication

Communication inevitably also occupies a key position in projects. Its goal is fast, open, correct and comprehensive information, because this is the raw material of knowledge.[40] The first axiom of communication by Paul Watzlawick is: "You cannot not communicate."[41] This means that not only words, but also facial expressions and gestures are communication.

Inadequate communication can have many negative effects. If no or too little information is given, the suspicion of concealment arises, and this is also a breeding ground for rumours. If you react too late, you leave the impression that you are not in control of the situation. If you spread different information on the same subject, you do not appear

[40] See Henn and Meyhöfer (2003).

[41] See Watzlawick et al. (1969), p. 53.

trustworthy. If you spread information as true that is demonstrably and known to be false, the speaker can quickly be exposed as a liar. The maxim must therefore be: One voice to the business—everyone learns the same thing at the same time. This reduces the risk of building a bad image—because such an image can threaten the very existence of a company: Less trust means fewer customers and less support from authorities, politics and banks.

5.6.1 Structural Measures for Communication

Space can have a huge impact on the quality, function and speed of work, and can also have a significant impact on the cost of the work done.

Building design can influence new ways of working such as planned communication, ensure user control and comfort, and increase functionalities. Before the user requirements for a building are formulated, the appropriate or desired design measures must be functionally described. To this end, all stakeholders are invited to an ideation and feasibility session, the goal of which is to describe downstream tasks, process change, building design, and integration of improved communication. If the user requirements are determined by communication, the conceptual design is reflected in a corresponding layout that promotes the encounter and to meet employees with each other which will encourage the communication.

An example of this is the construction of Columbia University Medical Center's Medical and Graduate Education Building, cooperation by the architectural firms Diller Scofidio + Renfro and Gensler. The 14-story building houses the university's dentistry, Mailman School of Public Health, and biomedical departments.

The "Study Cascade", descripes an open and for that a transparent staircase along the south façade, it provides spaces for study and social engagement between the different disciplines. There is no doubt that the, whicharrangement of the spaces is not occurred by accident, but rather by the layout which adapts the processes and requirements that should taking place in the building (Fig. 5.7).

5.6.2 Informal Communication

Application of network education is the main purpose of social contacts. In general, direct contact is considered the most efficient and sustainable exchange of information. Through constructive measures, informal communication can be promoted in planning and construction. It can enable the target of informal communication.

Fig. 5.7 Education Building at Columbia University Medical Center, USA

In accordance to the Allan-curve to which communication decreases the further away the communicators are from each other,[42] That has been known since the 1980s. It is taken into account that in office building planning, much more knowledge is available instead for planning of production buildings. In other words the knowledge gained from office planning with regard to health aspects has not yet been reflected in industrial buildings.[43]

The 3D arrangement of the obstructed elements contributes to the health of the employees and can promote informal communication. In addition to the flow of materials and personnel, the flow of communication must also be taken into account, which is equally essential for logistics as for informal and formal communication.[44]

The layout behaves proportionally to the spaces it "limits", especially when designed to meet needs. Thus, the space influences the work that takes place there and vice versa. This relationship is very easy to understand if you consider that you behave very differently in a church than, for example, in a discotheque.

The four most common ways in which the communication flow face-to-face can be designed as a sustainable communication are the following constructive designs (see Fig. 5.8):

- Junction
- Spine
- Atrium
- Cross coat

When planning meeting places or multi-purpose rooms such as a café or a small tea kitchen. It is necessary to define what function these rooms should fulfil and what type of communication flow would be most suitable for this. The question must be answered as to whether they should only be accessible internally or also publicly for everyone, e.g. in front of the plant barriers, whether the construction of connecting bridges makes sense, etc.

It can therefore be stated that a basic layout always starts from the structured user requirement. It must also be considered how the possible flows such as material, personnel and communication flows generally relate to each other (Fig. 5.9).

In production, the layout is mainly determined by the process. Here, the motto was to produce as much as possible. However, this has changed.[45] Nowadays, the goal is rather to produce the planned production quantities in consistent quality. To put it succinctly, this requires the well-being of the employees. An influencing part is the layout with the integration of staircases, lifts, kitchenettes and lounges and common rooms that invite informal communication and are points of social exchange.

---

[42] See Allen (1984), p. 241.

[43] See Lechler (1997).

[44] See Brandau (1985), p. 18.

[45] See Lechler (1997).

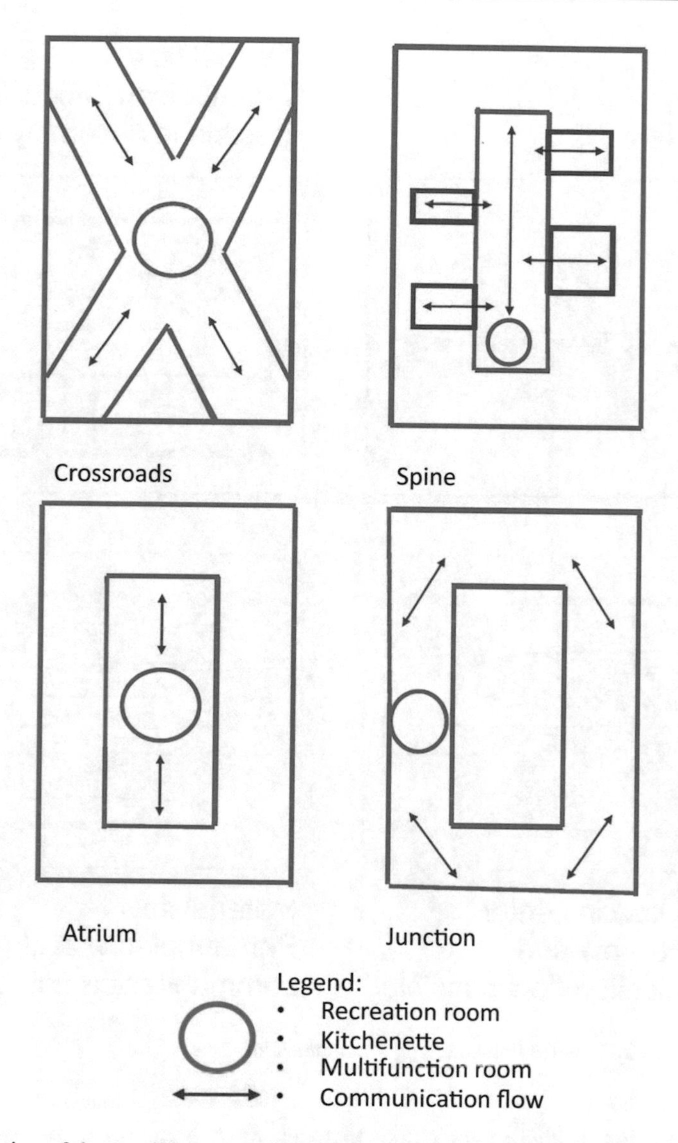

Fig. 5.8 Selection of the possible arrangement of the communication flows

Stairwells should be at the centre of information and communication flows, rather than in a remote corner, to encourage their use, which promotes movement and therefore health. Furthermore, the connections to the other departments and different floors should be structurally implemented in such a way that they invite people to stop and stay in order to exchange information with colleagues or superiors.

This also applies to kitchenettes and common rooms, which can no longer be accommodated in gloomy, possibly windowless rooms, because this lacks any appreciation for the employees. The solution is an open, centrally accessible and well-equipped

Material flow

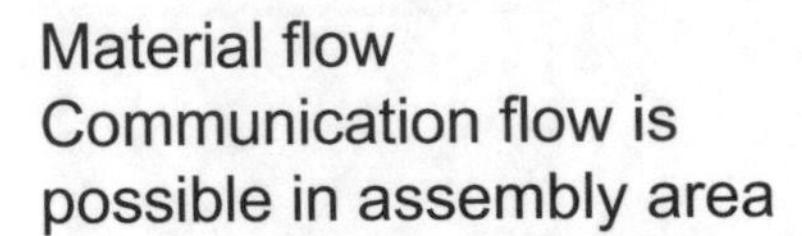

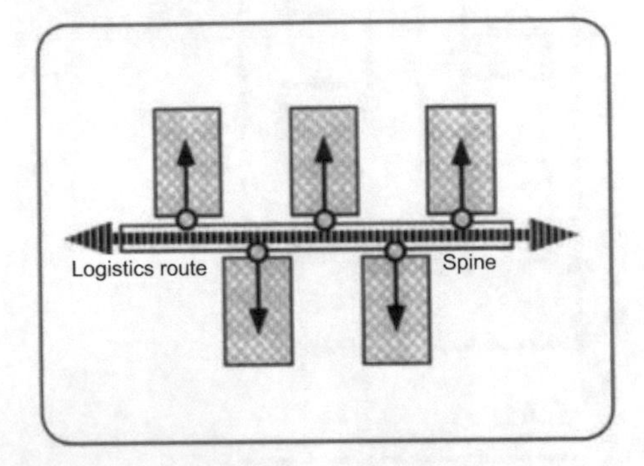

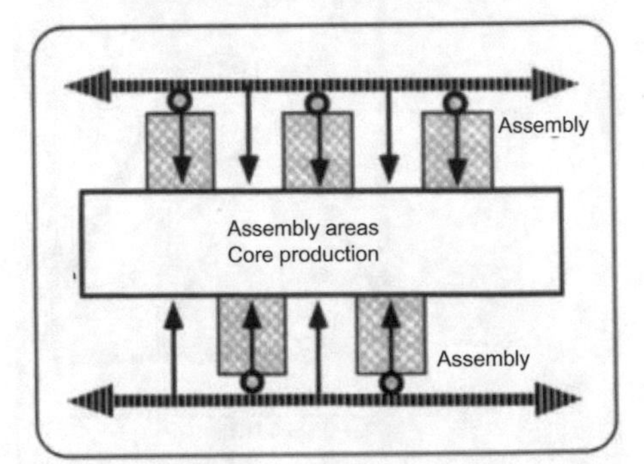

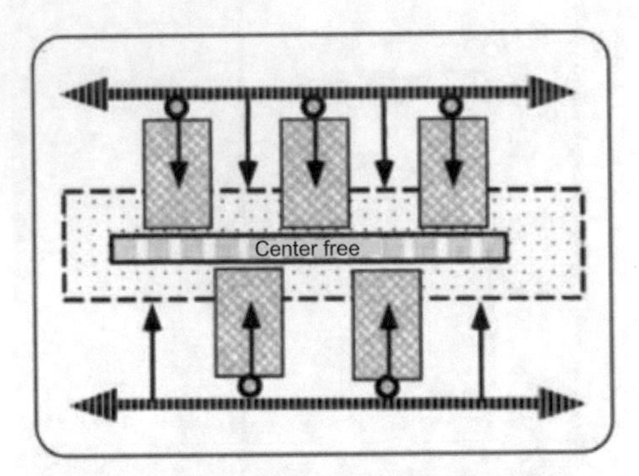

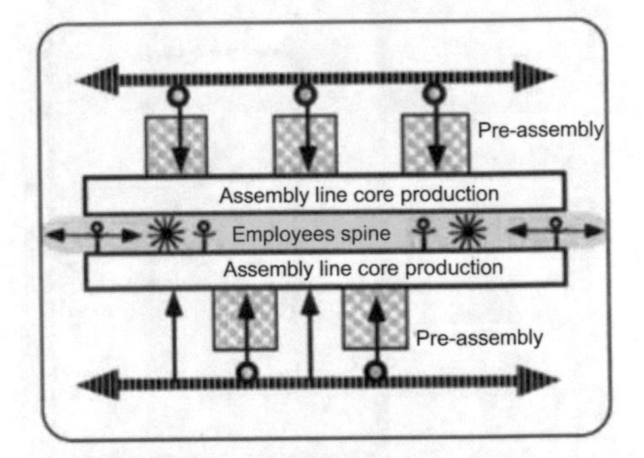

Material flow in center
Personnel flow and
communication flow possible

Material flow
Personnel flow and
communication flow possible

Fig. 5.9 Various options for implementing communication flows

multifunctional room where people like to meet. Such a room can be used to receive visitors or hold meetings, but can also be available for quiet work or a private phone call during quiet periods. Another effect of the central multifunctional room is to promote informal communication when all employees across the hierarchy levels have their coffee here.

It would also be conceivable to set up an area for public viewing, where presentations can be presented, successes celebrated or employees honoured.

If the walls between production, manufacturing or sales and the typing desk are removed, this strengthens the "we-feeling"of the employees because process steps

Fig. 5.10 Open relationships to promote communication

Fig. 5.11 Open structures in production

suddenly become visible and both customers and employees experience openness and transparency. This can also be achieved through the realisation of "non-territorial offices", the multiple occupation of a workplace by shift teams (Figs. 5.10 and 5.11).

5.6.3 Communication Concept

In many of the factors which already have described, aspects can be found that have to do with or promote communication. Communication serves not only to convey information or knowledge, but also very essentially to facilitate social interaction.

In order to bundle the individual measures in a targeted manner, it makes sense to define a communication concept depending on the situation and benefits. The ultimate goal of any communication concept is to structure the contact and exchange between management, employees and customers in order to contribute to the well-being of all. In every interaction, whether among employees, with superiors or subordinates, or with customers, a communication concept helps to maintain certain standards of social interaction and also to master challenging situations.

Such a concept includes the development of a corporate identity, which formulates the self-image of a company with regard to its goals, its working methods and its appearance and also takes into account appropriate wording. In addition, there is an Employers & Company Alignment, with which the company and its employees align themselves to a common vision, especially when change processes are pending.[46] The goal here is to establish a sense of "we", which leads to employees identifying with their company, being motivated and committed to contributing their skills and fulfilling customer requirements. These goals are supported by appreciative communication, the creation of inspiring work environments and the promotion of social exchange.[47]

5.7 Health

5.7.1 Hygiene and Allergies

The mutual agreement already describes above should also include general office rules for hygiene for the benefit of all. These include, for example, regular thorough cleaning of the desks, emptying the rubbish bins and providing disinfectant wipes so that you can also wipe the phone yourself once in a while. In addition, germ-resistant coated surfaces can also be considered.

Eating meals at the desk should be avoided; instead, a central meeting place in the office makes sense, where employees can drink coffee, eat something and exchange ideas or news with colleagues in between. There should be a dishwasher for used dishes. It is also important to regulate what can be stored in the refrigerator and for how long, so that spoiled food does not become a health risk.

[46] See Sect. 5.3.2.

[47] See Sects. 5.3.3, 5.6.1 and 5.6.2.

Regular hand hygiene should be a matter of course, soap and disinfectant must be available in the washrooms, and depending on the work area, showers are also useful. In addition, a visually good design of the sanitary areas promotes hygiene in the company and also shows the employee that he is important and valued.

Work areas should be ventilated several times a day at set times to reduce the number of viruses and bacteria in the room through air exchange. This can also help prevent allergies. An allergic reaction in the office can have many causes. One of them is dust, so the cause of dust exposure needs to be found. Often it is the mix of pollutants, allergens and germs that actually make rooms unhealthy places. Sensitive people react to this with allergy-like symptoms, such as sick building syndrome. Dry air can also irritate the respiratory tract.

5.7.2 Place of Silence/Power Napping

If you ask people in confidence about their place of silence at work, often comes the following answer: the toilet. This can hardly have been meant by modern workplace situation—other, separate rooms are needed for this.[48] There are several options for what such a place of rest might look like:

Power napping doesn't just make up for a sleep deficit. According to new studies, it also strengthens the immune system and lowers stress levels. In a power napping station, you can recover within a very short time. After only 15 to 20 min, the recovery sets in noticeably. As part of change management, it should be agreed that every employee may use this station. This also requires the participation of the supervisor, who sets a good example.

The Activity-Based Workplace is the place to be if you want to work in peace and quiet, whether in a small meeting room with an armchair or at a desk in the quiet zone. If there is no window with a view, a picture is a good substitute.

If a garden, park or forest is within reach as a place of retreat, a short walk can serve as an inner retreat. More and more planners are envisaging a large garden or park for hotels or public buildings, such as at the new DFB performance centre in Frankfurt/Main.[49]

5.7.3 Mental Activity

The urge to engage in a demanding activity is often only a matter of time, once one has indulged in recreation long enough on vacation, for example. Among the offers of mental activity in the leisure sector is the library on a cruise ship as well as the open-air chess facility in the hotel garden.

---

[48] See Mühl (2019).

[49] See Ashelm (2019).

Since demanding work is already being done at all hierarchical levels in the commercial sector, the pressure for intellectual activity is quite low. Here, however, offers such as a company choir or orchestra could create a balance and set other impulses; moreover, they would also serve to establish a sense of togetherness.

Boredom sets in quickly, especially in waiting areas of all kinds. In order to stimulate more intellectual activity, for example, specialist magazines, popular science magazines or brain games could be displayed in surgeries in addition to the usual lifestyle and social magazines.

5.7.4 Safety/Quality

The feeling of safety has a fundamental influence on people's emotional state and sense of well-being. Safety implies protection against unforeseeable events or the effects of incorrect operation of equipment, in that it can be assumed that something has been installed according to regulations and that its use follows legal requirements, user requirements, a manual or similar.

In order for this confidence in safety to be justified, it is necessary to implement quality management and quality assurance, which is regularly certified by independent examination and analysis in accordance with qualification.

5.7.5 Sport and Exercise

In principle, commercial employees in their defined sphere of activity do not suffer from an acute lack of exercise. They are not among the 1.4 billion people in the world who move too little. After all, around a third of women and a quarter of men are affected by acute lack of exercise.[50]

Movement at the workplace, like some other factors, can be promoted by constructive and organisational measures, e.g. by the location of lifts and staircases or by placing the printer, wastepaper bin, etc. in a central position in the open space.

Instead of calling or even emailing the colleague sitting around the corner, a short visit can be paid. Some meetings could also be held standing or walking. To this end, it makes sense to equip some meeting rooms or parts of the canteen with standing tables. Activity-Based Working can be supported with mats, exercise chairs and possibly rings to hang from. Some companies already offer sports rooms for fitness, yoga or back training, which are equipped with devices for more targeted training.

[50] See Watts et al. (2018), p. 8.

Breaks (especially the lunch break) can also be used as exercise breaks. Getting out of the office and into the fresh air is not only good for the musculoskeletal system, but also for the mind.

Outside the workplace, everyone can ensure more exercise themselves by stopping at public transport or—especially in good weather—cycling to work; if the route is too long, train and bike can also be combined. The employer can support this by providing bicycle parking facilities in sufficient number and quality.

The motivation to implement the aforementioned behavioural tips can be promoted by company sports groups and participation in company runs or regional sports tournaments—and this also serves to strengthen the sense of unity.

References

AIRY Greentech GmbH (2020), Unsere Technologie, https://airy.green/de/technology [Zugriff am 29.05.2020].

Aldwin, C. M. (2007), Stress, coping, and development – An integrative perspective, Guilford, New York.

Allen, T. J. (1984), Managing the flow of technology: technology transfer and the dissemination of technological information within the R & D organization, MIT Press, Cambridge (Mass.) / London.

Ashelm, M. (2019), Der DFB und das Silicon Valley des Fußballs, in: Frankfurter Allgemeine Zeitung, 03.05.2019.

Bachmann, P. (2018), Bedarfsgerechte Gebäudetechnik, in: Seiferlein, W., Kohlert, C. (2018), Die vernetzten gesundheitsrelevanten Faktoren für Bürogebäude, Springer Vieweg, Wiesbaden, S. 139–150.

Braem, H. (2009), Die Macht der Farben: Bedeutung & Symbolik, Wirtschaftsverlag LangenMüller/ Herbig, München.

Brämer, R. (2008), Grün tut uns gut – Daten und Fakten zur Renaturierung des Hightech-Menschen, natursoziologie.de, 5/2008, https://www.wanderforschung.de/files/gruentutgut1258032289.pdf [Zugriff am 11.06.2020].

Brandau, R. (1985), Pharmaproduktion aus betriebswirtschaftlicher Sicht, Wissenschaftliche Verlagsgesellschaft, Stuttgart.

Bundesverband Bürohund e. V. (2019), Bürohund-Index Deutschland 2019, http://bv-bürohund.de/ buerohund-index-deutschland-2019-ergebnisse/ [Zugriff am 08.06.2020].

Graen, A. (2019), Sieben erwiesene Gründe, warum in jedem Büro ein Hund sein sollte, in: stern, 27.06.2019, https://www.stern.de/neon/vorankommen/karriere/7-gruende–warum-in-jedem-buero ein hund sein-sollte-8772744.html [Zugriff am 80.06.2020].

Haapakangas, A., Kankkunen, E., Hongisto, V., Virjonen, P., Oliva, D., Keskinen E. (2011), Effects of Five Speech Masking Sounds on Performance and Acoustic Satisfaction. Implications for Open-Plan Offices, in: Acta Acustica united with Acustica, 97/2011, S. 641–655.

Hahn, N. (2007), Der Einfluss von trockener Luft auf die Gesundheit, in: Gefahrstoffe – Reinhaltung der Luft, 3/2007, S. 103–107.

Heidari, A. A., Tavakol, H., Behdadfar, N. (2013), Effect of Lighting and space on depression and stress appearing in residential places, in: Journal of Novel Applied Sciences, 12/2013, S. 733–741.

Henn, G., Meyhöfer, D. (Hrsg.) (2003), Architektur des Wissens – Architecture of Knowledge, Junius, Hamburg.

Hortig, J. (2013), Das Geschäft mit dem Geruch – Wenn der Zug nach Rosenholz duftet, https://www.wiwo.de/unternehmen/handel/das-geschaeft-mit-dem-geruch-wenn-der-zug-nach-rosenholz-duftet/8464296.html [Zugriff am 11.06.2020].

Kals, U. (2019), Kuschelfaktor im Büro, Interview mit J. Brettmeister, in: Frankfurter Allgemeine Zeitung, 07.07.2019.

Kasper GmbH (2020), Digitaldruck, http://www.kasper-werbung.de/digitaldruck-singen.html [Zugriff am 29.05.2020].

Kutchma, T. M. (2003), The Effects of Room Color on Stress Perception: Red versus Green Environments, in: Journal of Undergraduate Research at Minnesota State University, Mankato, 3/2003, Article 3.

Lechler, T. (1997), Erfolgsfaktoren des Projektmanagements, Peter Lang, Frankfurt am Main.

Mai, J. (2016), Wertschätzung: Mehr als Belohnung und Lob, https://karrierebibel.de/wertschaetzung/#Die-Ausdrucksformen-der-Wertschaetzung [Zugriff am 29.05.2020].

Martin, M., Schlipat, H. (2014), Wertschöpfung durch Wertschätzung – Warum das „gesunde Unternehmen" wesentliches Element der Unternehmensstrategie sein sollte, https://www.rochusmummert.com/downloads/news/Wertschoepfung_durch_Wertschaetzung.pdf [Zugriff am 29.05.2020].

McCoy, J. M., Evans, G. W. (2002), The Potential Role of the Physical Environment in Fostering Creativity, in: Creativity Research Journal 14(3), S. 409–426.

Mehta, R., Zhu, R., Cheema, A. (2012), Is Noise Always Bad? Exploring the Effects of Ambient Noise on Creative Cognition, in: Journal of Consumer Research, 39/2012, S. 784–799.

Meyers-Levy, J., Zhu, R. (2007), The Influence of Ceiling Height: The Effect of Priming on the Type of Processing That People Use, in: Journal of Consumer Research, 34/2007, S. 174–199.

Mühl, M. (2019), Die revolutionäre Kraft der Achtsamkeit, Frankfurter Allgemeine Zeitung, 23.12.2019.

Nieuwenhuis, M., Knight, C., Postmes, T., Haslam, S. A. (2014), The Relative Benefits of Green Versus Lean Office Space: Three Field Experiments, in: Journal of Experimental Psychology: Applied, 20(3), S. 199–214..

Nüchterlein, P., Richter, P. G. (2008), Raum und Farbe, in: Richter, P. G. (Hrsg.), Architekturpsychologie, Pabst, Lengerich, S. 209–231.

Oetter, C. (2015), ISPE Fachgespräch, Leipzig, 24. Februar 2015.

Roelofsen, P. (2008), Performance loss in open-plan offices due to noise by speech, in: Journal of Facilities Management, 6/3, S. 202–211.

Rumpfhuber, A. (2013), Architektur immaterieller Arbeit, Tuna + Kant, Wien.

Seiferlein, W., Woyczyk, R. (2017), Projekterfolg – Die vernetzten Faktoren von Investitionsprojekten, Fraunhofer Verlag, Stuttgart.

Seiferlein, W., Kohlert, C. (2018), Die vernetzten gesundheitsrelevanten Faktoren für Bürogebäude, Springer Vieweg, Wiesbaden.

senkonzept GmbH (2017), Unsere Düfte mit biologischem Geruchsvernichter, http://www.senkonzept.de/duft.html [Zugriff am 08.06.2020].

Steelcase (2013), Wohlbefinden – ein Thema, das nur Gewinner kennt, 360° Magazin Nr. 8, https://www.steelcase.com/content/uploads/sites/2/2018/08/360N8DE.pdf [Zugriff am 08.06.2020].

Steelcase (2019), New Work. New Rules, 360° Magazin Nr. 16, https://www.steelcase.com/content/uploads/sites/2/2019/07/1023893hy_DE_MOCKUPsm.pdf

Stefani, O. (2017), Die Wirksamkeit von Tageslichtreplikationen auf die Wissensarbeit, Dissertation, Fakultät Maschinenwesen der TU Dresden.

Stone, N. J., English, A. J. (1998), Task Type, Posters, and Workspace Color on Mood, Satisfaction, and Performance, in: Journal of Environmental Psychology, 2/1998, S. 175–185.

Trautwein, K. (2018), Farbkonzepte für Arbeitsplätze, in: Seiferlein, W., Kohlert, C., Die vernetzten gesundheitsrelevanten Faktoren für Bürogebäude, Springer Vieweg, Wiesbaden, S. 102–112.

Watts, N., Amann, M., Arnell, N., Ayeb-Karlsson, S. et. al. (2018), The Lancet Countdown on health and climate change: shaping the health of nations for centuries to come, in: The Lancet, 392(10163), S. 2479–2514.

Watzlawick, P., Beavin, J. H., Jackson, Don D. (1969), Menschliche Kommunikation – Formen, Störungen, Paradoxien, Huber, Bern 1969.

6 How the Matrix Works

Preliminary Remarks
The purpose of this target matrix is to find out the most important well-being factors for different industries and areas. furthermore to determine to what extent they are applicable in order to create an individual concept from them. The target matrix has been pretested, for two use cases. These cases should be as different as possible The aim was to explore the performance limits for the application of the target matrix. The selected use cases were firstly a fitness center and secondly production in the pharmaceutical clean room sector.

6.1 Application 1: Fitness Centre

First, the current state was surveyed to determine which factors could be described as positive:

- The fitness center is very quiet
- It's easy to focus on what you're doing
- The occupancy is always pleasant
- No one needs to feel like they are without support, because there are highly trained staff to assist them

The fitness center has a standardized user requirement that is similar in some respects to the target matrix. In the following, the relevance of each factor is discussed, whereby only those factors are considered that are also feasible with regard to the present user requirement:

W. Seiferlein, *Well-Being Factors for Different Industries*,
https://doi.org/10.1007/978-3-658-34997-4_6

Architecture

- Healthy building and maintenance: *relevant*, although the studio is located in an existing building
- Functional room layout: *relevant*, concerns the placement of the devices in the room
- A parking space for each vehicle: *relevant*, this factor is already taken for granted

Ergonomics

- Adequate furniture: *relevant*, concerns lockers and the reception desk
- Office and workshop furniture: *not relevant*
- IT equipment: *not relevant*

Work–life Balance

- Mutual agreement: *not relevant*
- Employers & Company Alignment/Change Management: *not relevant*
- Appreciation: *relevant*, concerns the way customers are treated and the way employees behave towards each other

Perception

- Air: *relevant*, controlled inflow of fresh air or stove ventilation is desired by the user
- Light: *relevant* because adequate light is important in any room, with natural light preferred over artificial
- Noise: relevant, concerns the prevailing acoustics in the room, i.e. whether the room is reverberant or too heavily equipped with sound absorbers, so that the sounds of fellow trainees and possibly background music are too much or too little audible
- Body: *relevan*t, concerns offering water dispensers and possibly a vending machine with healthy goods such as fruit for customers
- Colors: *not relevant*, because currently the standard is white, the studio therefore looks very puristic, but a little color would not hurt
- Scents: *relevant*, concerns the smells of the equipment and exhalations of the fellow trainees. Covering rooms with scents is already standard, especially in the hotel industry

Nature

- Plants: *not relevant*, as they are not desired in the operator's standard requirements
- Water/sea: *not relevant*, as this is not included in the operator's standard requirements; however, it would be possible to integrate this point into a future concept and to hang up pictures
- Animals: *not relevant*

Communication

- Constructive measures: *not relevant*
- Informal communication: *not relevant*
- Communication concept: *relevant*, concerns communication with customers and promotional measures

Health

- Hygiene/allergies: *relevant* because their implementation represents a valid need in a fitness centre
- Safety/quality: *relevant*, concerns the functioning of the equipment
- Place of rest/power napping: *not relevant*
- Intellectual activity: *not relevant*
- Sport and exercise: *relevant* because this is the business purpose of a fitness centre

In order to visualize the result of the relevant factors, the target matrix is subsequently adapted to the result of the evaluation of the factors (Fig. 6.1).

6.2 Application 2: Pharmaceutical Production

The relevant well-being factors were also identified for this use case:

Architecture

- Healthy construction and maintenance: *relevant*, as there is no standard, the choice for construction and materials has to be made individually
- Functional room layout: *relevan*t, especially if material, personnel and communication flows are to be created
- One parking space for each vehicle: *relevant*, shortens the journey to and from work for employees

Ergonomics

- Appropriate furniture: *relevant* because it is important to use the appropriate furniture depending on the activity, which can perhaps be called “Activity-Based Furniture” (see above)
- Office and workshop furniture: *relevant* (see above)
- IT equipment: *relevant* (see above)

Architecture	Healthy building and maintenance	Functional room layout	A parking space for each vehicle			
Ergonomics	Adequate furniture	~~Office and workshop furniture~~	~~IT equipment~~			
Work-life balance	~~Mutual agreement~~	~~Employers & Company Alignment/ Change Management~~	Appreciation			
Perception	Air	Light	Noise / music	Body	Colors	Fragrances
Nature	~~Plants~~	~~Water/ sea~~	Animals			
Communi-cation	~~Constructive measures~~	~~Informal communi-cation~~	Communi-cation concept			
Health	Hygiene/ allergies	Safety/ quality	~~Place of rest/ power napping~~	~~mental activity~~	Sport and movement	

Fig. 6.1 Target matrix adapted to the fitness centre

Work–life Balance

- Mutual agreement: *relevant*, especially if an intensive change is planned
- Employers & Company Alignment/Change Management: *relevan*t, particularly in the planning phase especially in the case of changes to the company structure and in connection with personnel
- Appreciation: *relevant* because the different factors of when a person feels good have a lot to do with the issue of appreciation

Perception

- Air: *relevant*, concerns the section before loading or filling process where the product is still handled "open", up to a controlled (clean room) area
- Light: *relevant*, "Human Dynamic Light" is to be used according to human needs, as far as it is technically possible
- Noise: *relevant*, as continuous exposure to sound above the limit value is not permitted; sound sources may have to be relocated or encapsulated
- Body: *relevant*, concerns the food from the canteen and the contents of the vending machines (fruit instead of chocolate bars)
- Colours: *relevant*, this has been shown by the colour design of the production facilities by von Garnier
- Fragrances: *not relevant*, as there is too much inherent odour in the production process and a mixture of fragrances would not be effective

Nature

- Plants: *not relevant* because of the risk of contamination
- Water/sea: *relevant* in areas where pictures can be displayed and a landscape wallpaper can be brought in
- Animals: *not relevant*, because there is a risk of contamination and the owners cannot look after the animals due to the production process. Only in the break area an aquarium or a screen with fish, a campfire or similar would be conceivable

Communication

- Constructive measures: *relevant* to promote the flow of communication
- Informal communication: *relevant*
- Communication concept: *relevant* in order to implement all measures in terms of planning

Health

- Hygiene/allergies: *relevant*
- Safety/quality *relevant*: under no circumstances should compromises be made here; in this area the motto "safety first, quality always" always applies
- Place of rest/power napping: *relevant*, this maintains staff performance throughout the day
- Intellectual activity: *not relevant* because this is an essential part of the work and does not need to be encouraged
- Sport and exercise: *not relevant* because this is part of the daily work and does not need to be promoted

The target matrix was also adapted to the requirements for this application (Fig. 6.2).

6.3 Evaluation of the Use Cases

The corresponding well-being factors were determined for the two use cases. It became apparent how flexible the matrix can be used even for very different areas.

In the following step, the matrix is coordinated in dialogue with the users and operators. Once all well-being factors have been defined, an overall concept is created. For its implementation, the necessary acceptance should be achieved for the defined parameters of the matrix. This is done with active communication in the sense of change management, so that the advantages and disadvantages can be weighed up jointly with all those involved and accepted or rejected.

Architecture	Healthy building and maintenance	Functional room layout	A parking space for each vehicle			
Ergonomics	Adequate furniture	Office and workshop furniture	IT equipment			
Work-life balance	~~Mutual agreement~~	~~Employers & Company Alignment/ Change Management~~	Appreciation			
Perception	Air	Light	Noise / music	Body	Colors	~~Fragrances~~
Nature	~~Plants~~	Water/ sea	Animals			
Communi-cation	Constructive measures	Informal communi-cation	Communi-cation concept			
Health	Hygiene/ allergies	Safety/ quality	Place of rest/ power napping	~~mental activity~~	~~Sport and movement~~	

Fig. 6.2 Target matrix adapted to pharmaceutical production

Summary

7

Preliminary Remarks
With the application of the well-being factors, the performance of the employees increases due to an optimization and improvement of the physical, mental and social components induced by them, which leads to less absentia and fluctuation. Another aspect is employee engagement, which increases as a function of job satisfaction. The effects on the workplace caused by the implementation of well-being factors are manifold and visible. As these soft factors are not measurable, they are also difficult to compare.

The research has shown that numerous industries or sectors already deal with well-being factors, but do not always call them with the same. For example, the terms "comfort" or "ambience" are used in cruise shipping, and the term "feel-good atmosphere"is used by Deutsche Bahn, indicating knowledge of these factors or the processes they trigger. However, the term "well-being factors" is already frequently used in current conversions and new buildings in the pharmaceutical sector.

Often the effect of these factors is not known, and even more rarely are the possible well-being factors identified and concretised within the framework of a concept. The extent to which such a concept is created often depends on the sectors and areas.

Each industry focuses on different details and thus requirements. The aim of each planning is to elicit the requirements in an expedient and user-oriented manner and to integrate them into the catalogue of requirements. As a result, experts from different disciplines are appointed to the planning teams. With the interdisciplinary groups, different perspectives and opinions are captured, which expand the finding and use of the well-being factors.

W. Seiferlein, *Well-Being Factors for Different Industries*,
https://doi.org/10.1007/978-3-658-34997-4_7

> The classic models will soon be obsolete. In the future, there will no longer be a distinction between blue-collar and tie-wearing. An interdisciplinary and open working world will require a completely new way of thinking.[1]

For the layout of a work area, it is well known that the material and personnel flows to be taken into account must be planned. The layout also has a major influence on communication. By including communication needs at an early stage, it should also be possible to optimise the communication flow. In addition, it is important to integrate well-being factors and communication flows for the corresponding industries or areas in terms of planning and design.

Since workplaces are structured very differently in different industries and sectors, the aim of this book was to examine whether the well-being factors found for the office sector also apply to different occupational groups and whether they have the same effect on people with different occupational functions. It was initially assumed that the relevant factors would also be applicable to people in other sectors and industries. To verify this, essentially the statements of specialists in their field were collected in interviews and discussion groups and supplemented by a literature search.

Thus, 7 categories with 26 valid factors were identified and a target matrix was created from these. The individual factors are summarized in Chap. 4. The target matrix serves as a uniform summary and is relevant for all sectors.[2]

The individual factors found are described in Chap. 5, so that the target direction is recognisable. Since a comprehensive description of the factors could not be carried out, it merely serves as a guideline, which—as is quite intentional—gives room for interpretation.

At this point, the most outstanding and unusual ideas should be mentioned in particular.

In the shopping center section, a very interesting point was mentioned that was not found in any other literature, namely that parking is always available for both customers and employees.

The idea train of the Deutsche Bundesbahn comes up trumps with good ideas. This applies both to the constructive improvements ("standing seat"and "alternating seat") and to the public viewing, which has found its way into the well-being factor communication concept.

Surprising was the result from the horticulture topic, which has a very wide variety of factors.

No convincing piece of furniture has yet been found for the function of power napping from the field of health. The aim is to relax the eye, the body and the mind for a short time. A solution for a power napping station comes from the Deutsche Bundesbahn from the Ideenzug project.

This empirical investigation gives rise to questions that could not be dealt with here:

---

[1] Köhn (2017).

[2] See Appendix 1 and 2.

- Can and should a rating be applied to the factors?
- What is the functional effectiveness of the factors depending on the industries and fields?
- What about the universal effectiveness of the factors depending on people from different industries and fields?
- How is digitalization changing society? Do we have to question our existing model of society?
- What responsibilities will companies have to assume in the future?
- How do we make consistent use of the opportunities that present themselves and how do we succeed in taking advantage of them?
- Where is production heading in terms of human-machine collaboration?

The aim was to learn about and examine the relevant well-being factors from different industries and fields. Not all industries were considered, for example, it would be very interesting to examine the automotive industry for well-being factors.

The trend of the future is clearly moving in the direction of Industry 4.0, but the goal will no longer be digital planning and analogue construction, but will consist in a complete digital penetration of all process steps. To achieve this, all the skilled workers involved in a project must be able to use the digital tools and communicate with each other, and the training of tomorrow's skilled workers must take this into account. The professional exchange between the skilled trades and research will therefore become increasingly important, whereby the skilled trades can also be upgraded by digitalisation and the new opportunities it brings. In all sectors, well-trained professionals will be needed to collaborate with machines and technologies and to work efficiently and save resources, because "[a] human has experience and creativity, a machine precision and speed."[3]

References

Köhn, R. (2017), Blaumann und Krawattenträger an einem Tisch, in: Frankfurter Allgemeine Zeitung, 03.02.2017, S. 19.

Seiferlein, M. (2018), Zukunft des Bauens, in: Frankfurter Bauzeitung, 55/2018, S. 7–8.

[3] See Seiferlein (2018).

Appendix

Appendix 1: The Well-being Factors as a Checklist

Architecture

- Healthy building and maintenance
- Functional room layout
- One parking space for each vehicle

Ergonomics

- Adequate furniture
- Office and workshop furniture
- IT equipment

Work-life Balance

- Mutual agreement
- Employers & Company alignment/Change management
- Appreciation

Perception

- Air
- Light
- Noise
- Body
- Colors
- Scents

Nature

- Plants
- Water/Sea
- Animals

W. Seiferlein, *Well-Being Factors for Different Industries*,
https://doi.org/10.1007/978-3-658-34997-4

Communication

- Constructive measures
- Informal communication
- Communication concept

Health

- Hygiene/Allergies
- Safety/Quality
- Place of rest/Power napping
- Intellectual activity
- Sports and exercise

Appendix 2: Target Matrix for Visualising the Results

Architecture	A parking space for each vehicle	Healthy construction and maintenance	Functional room layout			
Ergonomics	Adequate furniture	Office and workshop furniture	IT equipment			
Worklife Balance	Mutual agreement	Employers & Company Alignment / Change Management	Appreciation			
Perception	Air	Light	Noise / Music	Body	Colours	Fragrances
Nature	Plants	Water and sea	Animals			
Communication	Structural measures for communi - cation	Informal communi - cation	Communicati on concept			
Health	Hygiene and allergies	Safety / quality	Place of rest/power napping	Mental activity	Sport / Moveme nt	

List of Photos

Despite of careful research, we were unable to locate all of the originators. These images are identified by the note "originator unknown", these were indispensable for the text. Please do not hestitate to contact us in this matter.

Originator	
p. 14	Sanofi
p. 15	Sanofi
p. 16	Merckle
p. 17	Werner Seiferlein
p. 18	Werner Seiferlein
p. 18 above	Merckle
p. 19 below	Merckle
p. 20	Orginator unknown
p. 21	Orginator unknown
p. 27	Werner Seiferlein
p. 28	Werner Seiferlein
p. 29	Werner Seiferlein
p. 30	Werner Seiferlein
p. 31	Werner Seiferlein
p. 32	Werner Seiferlein
p. 35	DB
p. 36 above	DB
p. 36 below	DB
p. 37	DB
p. 38	DB
p. 37	DB
p. 40	Meyer shipyard
p. 55 below	DB
p. 57	Meyer shipyard
p. 43	Dieburg Savings Bank

(continued)

W. Seiferlein, *Well-Being Factors for Different Industries*,
https://doi.org/10.1007/978-3-658-34997-4

Index

W. Seiferlein, *Well-Being Factors for Different Industries*,
https://doi.org/10.1007/978-3-658-34997-4

Printed in the United States
by Baker & Taylor Publisher Services